BASIC
BUTCHERING

BASIC BUTCHERING

——OF——

LIVESTOCK & GAME

JOHN J. METTLER, JR., D.V.M.

Illustrated by Elayne Sears

A Garden Way Publishing Book

STOREY COMMUNICATIONS, INC.
POWNAL, VERMONT 05261

Design and production by Andrea Gray
Photographs by Elinor Mettler

© **Copyright 1986 by Storey Communications, Inc.**

Typeset by Quad Left Graphics, Burlington, Vermont, in Aster.
Printed in the United States by Capital City Press.
Tenth Prin▋

Library of Congress Cataloging-in-Publication Data
Mettler, John J., Jr., 1923–
    Basic butchering of livestock and game.

    "A Garden Way Publishing book."
    Includes index.
    1. Slaughtering and slaughter-houses. 2. Meat
cutting. I. Title.
TS1960.M55   1985   664'.9   85-70195
ISBN 0-88266-391-7 (pbk.)
ISBN 0-88266-392-5

CONTENTS

ACKNOWLEDGMENTS

My wife, Elinor, collected all the recipes for this book and wrote the descriptions of the many cooks who contributed them. Editor and publisher of a weekly country newspaper, she also applied her editing skills to the manuscript.

Many people were consulted on practical methods of home butchering. Our friend, David Silvernail, allowed us to observe and photograph his annual hog butchering — from preparing the hogs and equipment, to salting and smoking the meat. His cousin, Jim Conklin of High Valley Farm, let us watch, photograph, and question while he slaughtered lambs and steers. Paul Dellea, a professional butcher, answered questions on everything from how to bone and roll a shoulder of lamb, to the best age for butchering a goat.

Scores of other people helped to write this book, often unknowingly. To them — my fellow country people, my friends and clients over the past forty years — I dedicate this book. Their ideas, if not their names, fill its pages.

PREFACE

WHEN I WAS A BOY everyone did their own butchering, either alone, as a family, or with a group of neighbors. I learned from hands-on experience, as my father and mother had learned from their parents. Today everyone is a specialist, butchering is done off the farm, and meat comes into the home wrapped in plastic and ready for the oven. The majority of a whole generation has never learned to butcher. Also today, however, many want to return to the subsistence farms, some even feeling that if we are to survive we must learn to supply more of our food ourselves.

As an army veterinarian in World War II, I was taught meat inspection and butchering techniques so I might teach people in occupied lands (the Pacific islands — Saipan in particular — in my case) to butcher hogs and cattle. More recently my butchering has been confined to a deer or two every year and an occasional duck or wild turkey. I have friends who butcher regularly, however, one a professional who cuts up meat for neighbors as well as in a local market, and others who butcher and process their own home-raised pork. They have been a tremendous help as technical advisers on this book.

You may want to be entirely self-sufficient on a small farm, or to avoid the high costs of the meat market, or to enjoy the better flavor of home-raised meat — or maybe you just want to butcher an occasional animal because you like being independent and doing things for yourself. In any case, this book is written to help you.

If you have enough do-it-yourself determination and a sufficiently mechanical mind to take things apart (in butchering you only take apart, you don't have to put back together) you can learn to butcher. Sure, you'll make mistakes, but they won't be life-threatening, and after a few trial-and-error experiences, you'll develop techniques of your own that will make you skilled in your own style. I will tell you one way to do things, but to paraphrase an old axiom, there is more than one way to cut up a hog.

INTRODUCTION

MAN HAS BEEN A MEAT EATER since the beginning of time, but he can also survive on nonmeat food such as grain, fruit, and vegetables. Some thinking people might ask, Why eat another animal when humans can survive on nonmeat food?

The answer is simple. Most of this planet grows plants — grass, trees, brush, moss, etc. — that man cannot eat. Only a small percentage of our land surface can efficiently grow plants that produce the high-quality protein our health requires (wheat, nuts, etc.). Ruminants — cattle, sheep, goats, deer — not only survive on grass and other herbage, they grow and thrive on them. The bacterial action of their rumen (paunch or first stomach) digests grass, leaves, moss, and small twigs into forms of protein and nutrients that the animals can absorb. Thus the protein in grass and other plants that man cannot digest is turned by ruminants into proteins (milk and meat) that he can eat and digest.

As for the hog, a nonruminant, he can grow and produce protein from farm waste — garbage, spoiled corn, garden refuse, sour milk — more products that man cannot eat. The chicken, although not as efficient a converter as the hog, can eat grain (corn) that is not easily digestible by man and produce high-quality protein in its meat and eggs.

THE LAW

The Federal Meat Inspection Act requires that all meat which is to be sold or traded for human consumption must be slaughtered under inspection in an approved facility under the supervision of a State or USDA meat inspector. A person can slaughter his animals outside such a facility only for use by him, members of his household, his nonpaying guests, and employees. He is not allowed to sell any portion of the carcass. For more details about these regulations, consult your county extension agent or write to the Food Safety Inspection Service, United States Department of Agriculture, Washington, D.C. 20250.

BASIC
BUTCHERING

1·TOOLS, EQUIPMENT, AND METHODS

A PROFESSIONAL BUTCHER, like a professional carpenter, has many specialized tools that make his work easier and help to make his finished job look better. This book is written for the do-it-yourself butcher who, like the do-it-yourself carpenter, wants to do a good serviceable job at the least cost. Special tools might make the finished product look better in the market, and so sell better; but we are interested only in how it looks and tastes on your dinner table.

Some are necessary, some are unnecessary but handy to have, and some are optional. For example, for veal or venison you only need one good all-purpose knife, such as a hunting knife, and a light rope, but a meat saw makes things easier. If you don't have a meat saw, a carpenter's saw will do.

For beef you need at least a couple of knives, preferably a skinning knife, a butcher knife, and a boning knife. You also should have a meat saw, but the job is made a lot easier with the use of an electric meat saw to split the carcass. For beef you do need some sort of lifting device: such as a block and tackle or come-along; rope, pulley, and tractor; or a single-tree. And you need a stunning hammer or gun and, of course, some good help.

Following is a list of equipment that is mentioned in the book. Where items are listed that you may not know by name, pictures or drawings are supplied. Some items have more than one use, such as the curved skinning knife used to skin beef and to stick hogs. Still, both jobs can be done with

1

something else, which is lucky because in the case of hogs, some of the butchering equipment is no longer made. For example, the double-edged sticking knife and the potash kettle are useful, but you'll only find them at an auction or antique shop. Many of the items listed, such as a tractor and scoop, are not essential, but if you have them handy, why not make use of them?

KNIVES AND ACCESSORIES:
    Skinning knife
    Butcher knife
    Boning knife (several styles available)
    Cleaver
    Bell scraper (for hogs)
    Sharpening stone
    Steel

SAWS:
    Hand meat saw
    Electric power meat saw (a similar saw made for wood
      is often used to split the backbone on beef)
    Electric band saw

LIFTING EQUIPMENT:
    Block and tackle
    Come-along
    Ropes
    Gambrel stick
    Singletree
    Tractor and pulley
    Tractor and scoop
    Hog hook or hay hook

OTHER EQUIPMENT:
    Axe
    Stunning hammer
    .22 single-shot rifle
    55-gallon drum, potash kettle, or 95-gallon stock tank
      to hold water to scald hogs
    Meat grinder
    Sausage stuffer

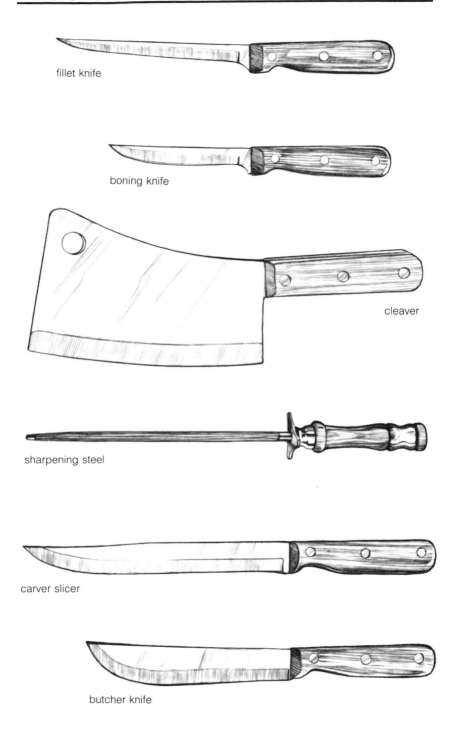

fillet knife

boning knife

cleaver

sharpening steel

carver slicer

butcher knife

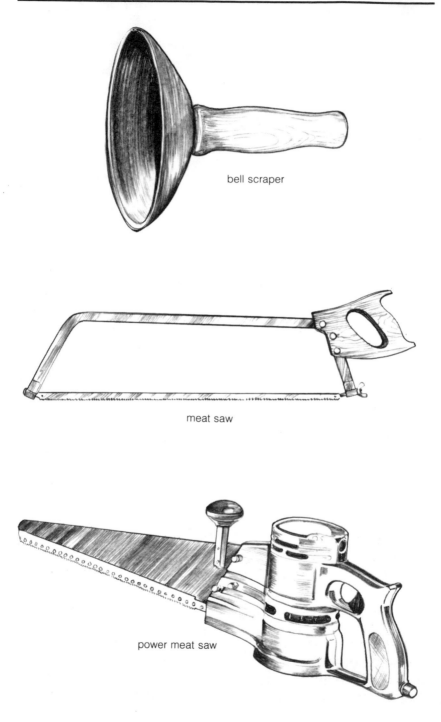

bell scraper

meat saw

power meat saw

Pails
Garden hose
Crocks for brine for salting (pottery or stoneware,
   wooden barrels, or heavy plastic garbage cans)
Floating thermometer
Hydrometer
Butcher string, freezer paper, and tape

SMOKEHOUSE:
Conventional, barrel, or small hobby size

More important than the variety of tools you have is how you use them. You can't do a good job of butchering without a sharp knife. Knives are sharpened on a stone at a 15- to 20-degree angle. Setting a sharpening stone in a frame made of ¼-inch wooden molding on a wooden work bench permits you to make full, even strokes from the heel of the knife to the point, and that allows even wear on the stone and on the knife. Most people can sharpen a knife on a Carborundum stone, or better yet use a Carborundum stone followed with an oil stone. However, you can't get a really fine edge on a knife without the use of a steel, probably the most misunderstood and most necessary piece of equipment the butcher uses.

When a skilled professional butcher uses a steel he touches the blade to the steel in a full stroke so lightly you can hardly hear the stroke. Slashing at the steel with a clatter, stroking so rapidly that the knife is a blur, is a sure sign of an amateur. To use a steel, hold it steady in one hand and with the knife in the other stroke ever so gently from the heel of the knife at the point of the steel to the point of the knife rubbing off at the base of the steel. Maintain the 15- to 20-degree angle of the blade to the steel as you did on the stone. Once a knife is sharp enough to shave with, even two or three strokes on each side every few minutes of work is all that is needed.

Meat saws can usually be sharpened by the same person who sharpens your hand carpenter's saws.

When you are through using butcher equipment, clean it all well, spraying metal parts with light oil if necessary, and store everything in a dry place. When you're ready to butcher again, spend a short time checking your equipment over, and wash it with detergent and hot water. Germicides for slaughter and butcher equipment are available, but for home use nothing beats detergent and plenty of hot water for getting rid of grease and dirt.

There are certain general observations that apply to slaughtering and butchering that are not necessary to repeat in the chapters dealing with specific animal species. For example, we can't influence the weather, but from experience you know when to expect the kind of weather you need for slaughtering. Keep weather in mind when you buy young pigs or as early as when you breed your ewes so as to have pigs or lambs ready for slaughter at the ideal time of year. Listen to long-range weather forecasts and pick a day to slaughter at the beginning of a period when daytime temperatures are 32° to 40° and nights go down to 25° but not much lower.

In order to butcher one must kill, but there is no need to be inhumane. The quality of the animal's life — that it has had sufficient food, clean shelter, and kind treatment from birth to death — is important; and death must come to all creatures at some time. The animal must be killed quickly, with little or no pain, but more important is that death comes without fear. To allow an animal to become frightened at slaughter is not only cruel, but unwise, for it causes the release of adrenaline, which some believe can affect the quality of the meat. Also, fear may cause the animal to struggle, doing damage to its meat or injuring the person slaughtering. Select the method of killing that will upset the animal's routine the least, thus avoiding fear, and select a method that is sudden, thus avoiding pain. Don't forget, however, that the heart must continue to pump briefly after death in order for the animal to bleed properly. Avoid sticking the heart of a pig, and don't delay cutting the throat of an animal that has been shot.

When you are cutting skin, cut from the inside out, particularly on deer and sheep, to avoid loosening cut pieces of hair or wool, which will give the meat a bad flavor.

One home butcher told me, "The secret of good-tasting meat is clean hands." Have soap, water, and paper towels handy while you are butchering. A garden hose long enough to use to spray a carcass is an excellent sanitary aid.

Construct your own cutting tables to heights that are comfortable for you to work at in a place that is cool and airy yet not drafty.

If you don't have access to a walk-in cooler to chill fresh carcasses, a double-walled (insulated) room with sliding doors to the north to allow you to regulate temperature is the next-best thing. When the weather is warm, close the door during the day and open it at night; when it is cool, close at night and open during the day.

Federal law prohibits sale of beef, veal, lamb, and pork not slaughtered at facilities under federal or state inspection. Thus you or your family are the ultimate consumer of your meat and the sole judge of how good a job you have done. As you cut, trim, and pack, keep that in mind so that packages are the correct size and cuts are those your family will enjoy most. If they like steak and stew better than roasts and hamburger, process the cuts accordingly.

Don't be afraid to improvise ways of cutting, tying, or preserving. Exchange ideas with other home butchers, listen to everyone, but do the job in the way that works best for you.

2·BEEF

ONE MIGHT SAY that until you have butchered a veal or lamb you shouldn't tackle a full-grown beef animal. Still, if you can round up at least one good helper and preferably two or three, and if you have the courage to try, you can butcher a beef easier than any other meat animal, on a pound-for-pound basis. If you have an experienced neighbor, you could be the helper when he butchers and then have him help you. Hands-on learning is the best way.

The Best Animal for Beef

The preferred beef animal is a steer at least 30 months old that has been confined to a stall or small pen and fattened on corn the last 30 to 60 days of its life. A heifer the same age will make nearly as good beef. In fact, a heifer 30 to 150 days pregnant will make better beef than either a steer or an open (unbred) heifer.

A cow or bull at least 30 months and up to about 5 years of age will make good beef if it has been confined to a stall and fattened for 60 to 90 days. Even after that the good quality of beef from certain individuals is amazing. A "short milker" (a cow that dries up too soon), pregnant about 90 days and fat, will make surprisingly good beef. In this hamburger age some farmers find that butchering an old cow and grinding everything but the most choice cuts is a good practice. However, don't grind beef until you've sampled a few cuts as roasts or steaks. I've seen cows as much as 11 years old that made good beef after standing in the barn all winter taking on fat.

Unless they are beef bred, animals under 24 months of age are usually half beef, half veal, and can be either tough or lacking flavor, or both.

Getting Ready

Unless you have a place to hang and age it, plan your beef butchering for late fall, late winter, or early spring. A clear day in the low 40s is perfect. Confine the animal to a small, clean pen and withhold feed for 24 hours. Allow access to water. It takes days to really starve out a full-grown beef, but 24 hours without hay or grain will reduce some volume and weight when you are removing the viscera of your beef.

Before you start to butcher, get all your tools together and ready in the butchering area, which should be clean and swept free of dust, cobwebs, and hay overhead that could fall while you are butchering and hanging your beef. You will need at least one sharp butchering knife, a skinning knife and an extra knife for each helper, with a steel and sharpening stone. For cutting up beef you will need at least one boning knife. A hand meat saw or electric power saw are handy, but a carpenter's saw can be substituted.

A stunning hammer or a gun is needed to kill the animal. Some people consider a .22 too small, but if properly used with long-rifle ammunition it is big enough. For lifting you will need a block and tackle, a come-along, or a rope and pulley and tractor to pull it. A heavy singletree is usually safer to use than a gambrel stick. A heavy 1½-inch pipe with a ring welded in the center to keep the lift from slipping sideways may also be used for a metal "gambrel stick." But unless the rear legs are wired or secured in some way, the carcass may slide off the pipe while you are splitting or quartering which could cause serious personal injury or loss of the meat to spoilage.

You should have pails, soap, and paper towels and, if possible, a water hose with water left running a bit to keep from

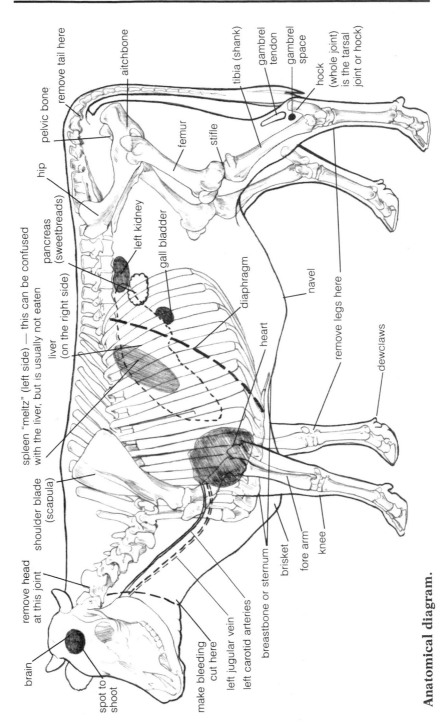

remove tail here

aitchbone

tibia (shank)

gambrel tendon

gambrel space

hock (whole joint) is the tarsal joint or hock)

pelvic bone

remove tail here

hip

femur

stifle

pancreas (sweetbreads)

left kidney

gall bladder

spleen "meltz" (left side) — this can be confused with the liver, but is usually not eaten

liver (on the right side)

diaphragm

navel

remove legs here

heart

dewclaws

shoulder blade (scapula)

remove head at this joint

brain

spot to shoot

make bleeding cut here

left jugular vein

left carotid arteries

breastbone or sternum

brisket

fore arm

knee

Anatomical diagram.

freezing if the temperature is below 30°. Containers for liver, heart, and sweetbreads, and a means of disposing of other insides, should be ready ahead of time. String to tie the animal's anus, or bung, should be handy when you need it. Light ropes, an extra rope halter, and an axe should be available but may not be needed.

Killing and Skinning

When all is ready, shoot or stun the animal, not by a blow between the eyes, but by striking above the eyes and just off center of the forehead **(figure 2.1)**. If you drew a line from each eye to the opposite horn, forming an X, the spot to aim for would be just beside where the lines crossed, as shown. For shooting cattle a shotgun at 12 to 18 inches is deadly and less dangerous to people and other livestock than a rifle or pistol. No matter how you kill the animal be as humane as possible.

The animal's throat should be cut immediately after the animal is shot or stunned. Cut just behind the jaw **(figure 2.2)**. Be sure to cut as deep as the bone to get not just the jugular vein but the carotid arteries as well. You can tell you've cut the arteries when you see and hear blood spurt.

If you have a tractor with a lift capable of safely picking up the animal, you can kill it right in the stall and then carry the dead animal by the hind legs to where you are going to butcher. A quiet animal, of course, may be led to the butchering area, or you can kill the animal and then drag it with a tractor or tackles to where you are going to butcher.

If you have done a good job of cutting both carotid arteries, the animal will bleed out well whether you hang it up or not. As soon as you are sure that the animal is dead, cut off the dewclaws of each rear leg and skin from there toward the body until you have uncovered the hocks **(figure 2.3)**. Hook the hocks from a singletree through the gambrel space **(figure 2.4)**. When the hocks have been carefully secured in the gambrel space you may remove the legs at the location shown in the anatomical diagram on page 11. Use a saw if you can't find the exact joint. Be careful not to cut the legs off too close

Figure 2.1: Shoot or stun the animal by a blow above the eyes, and just off the center of the forehead.

Figure 2.2: Cut the animal's throat immediately after it has been shot or stunned. Cut just behind the jaw and as deep as the bone in order to get the jugular veins and carotid arteries.

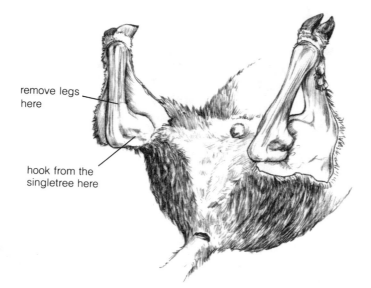

remove legs
here

hook from the
singletree here

Figure 2.3: Skin the rear legs from the dewclaws toward the body until you have uncovered the hocks.

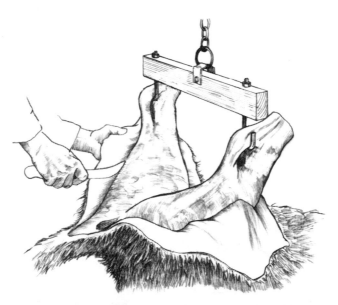

Figure 2.4: Hook the hocks from a singletree through the gambrel space and skin the rounds. After the hocks are secure in the gambrel space, remove the legs at the place shown in the anatomical diagram and in figure 2.3.

to the hock or the carcass may pull loose of the gambrel space and fall off the singletree. As you continue skinning start to raise the animal with a block and tackle, heavy come-along, or pulley and rope pulled by a tractor. If you have a big enough tractor and small enough animal, a chain attached to the scoop of a tractor will do the job most easily.

Remove the skin as you lift, cutting the skin from the inside out where possible to avoid getting cut pieces of hair on the meat. Have warm water and paper towels handy so you can wash your hands every time they get soiled and before you touch any meat. It has been said that the secret to good-tasting meat is clean hands, so it will be worth the effort to take cleanliness seriously.

Before you lift the carcass you should have cut around the bung and tied it off with a string so manure can't leak out **(figure 2.5)**. Also, while the animal is still touching the ground, you can expose the flesh over the breastbone and brisket and, with a saw, split the breastbone from the brisket to the rear of the breastbone, being careful not to cut too far back and puncture the stomach **(figure 2.6)**. Now lift the carcass a little and cut into the abdominal cavity just ahead of the pelvis (in the area just forward of where udder or testicles would be). Take your time and be very careful here not to puncture an intestine. As you reach the abdominal cavity air will rush in, and you will have more room to work. Lift the animal higher, drop more skin, and using your fingers as a guide, or with your fist holding the knife in and the blade out, use the heel of the knife to open the abdomen on the midline all the way to the rear part of the breastbone **(figure 2.7)**.

Raise the animal clear of the ground, remove the hide, and cut off the front legs. Leaving the hide on the forequarter until now protects the meat from getting dirty. Now reach up into the pelvis from the inside, find the bung, and pull it down. Dissect the bung — and all the insides you can — from their attachments, and literally roll the whole mass out onto the ground **(figure 2.8)**. Cut the paunch (rumen) free from where the gullet (esophagus) comes through the skirt (diaphragm). Or better yet, cut the diaphragm and remove the lungs and heart with the gullet still uncut. Remove the liver,

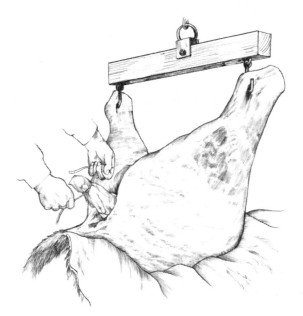

Figure 2.5: Dissect the bung and tie it off with a piece of string. Do not lift the carcass until you have done this.

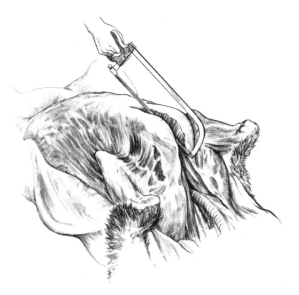

Figure 2.6: Expose the flesh over the breastbone and brisket. Then, with a saw, split the breastbone from the brisket to the rear of the breastbone. Be careful not to cut too far back or you may puncture the stomach.

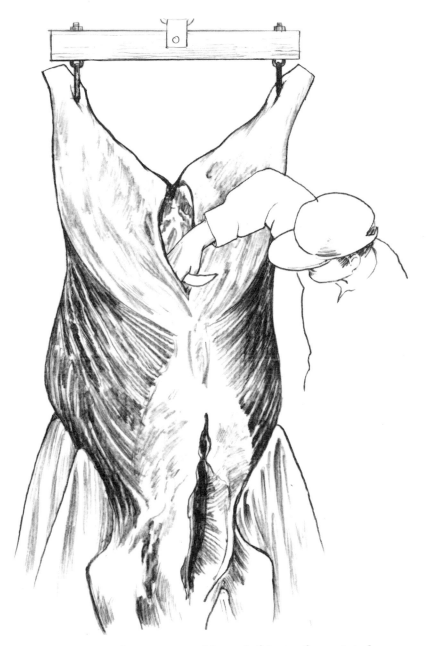

Figure 2.7: Raise the carcass a bit and skin to the point shown. Next, cut into the abdominal cavity just ahead of the pelvis. With your fist holding the knife in and the blade out, open the abdomen on the midline all the way to the rear part of the breastbone.

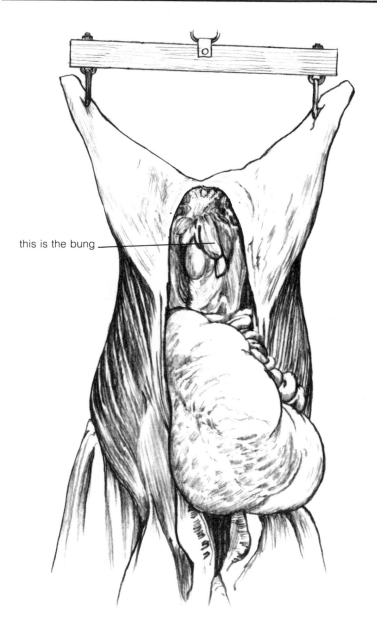

this is the bung

Figure 2.8: Reach up into the pelvis from the inside, find the bung, and pull it down into the open cavity. Dissect the bung and other visible organs from their attachments, and roll it all onto the ground.

cut out the gall bladder, and hang the liver to cool. If you can find the belly sweatbread (pancreas) under and behind the liver on the animal's right side, remove it and slide the rest of the insides out of the way so as to avoid tripping on them and falling.

Spread the breastbone with a stick and cut the skirt (diaphragm) close to the ribs and remove the "pluck" (heart and lungs). Save the heart. Open it to remove the blood clot, rinse, and hang to cool. Take the windpipe and gullet out all the way to where you cut them when you cut the animal's throat. Remove and save the tongue. Now remove the head at the first joint between the head and neck **(figure 2.9)**. If you wish, dissect the cheek meat away from the jaws. It is good meat for stew, or you can feed it to the dog.

If you are not sure you have removed all the debris from the pelvic canal, you may now split the pelvis with a saw on

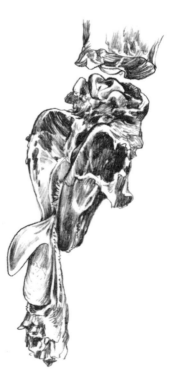

Figure 2.9: Remove the skinned head at the first joint between the head and the neck. Drop the tongue out of the head, remove it, and save it.

the midline in the area where the testicles or udder hung. This will make it easier to trim away visible nonedible material such as glands, portions of the penis, or tissue contaminated with dirt or manure. Now rinse the inside of the carcass with cold water from a hose or thrown from a pail to remove any loose blood clots or contamination from dirt or intestinal contents.

Quartering and Butchering

Split the carcass by sawing down the length of the backbone, leaving the two equal halves of the carcass joined by the last few inches of neck tissue **(figure 2.10)**. Splitting can be done with an axe or large cleaver, but that takes skill. Electric power saws are made for the job. If you can borrow one it will be a great help. In some areas where several people butcher, neighbors may buy one as a community property. You may also use a carpenter's power saw and, if you don't have a hand meat saw, you can substitute a carpenter's hand saw.

If you're lucky the temperature will remain between 32° and 40° so your beef can age properly for a week before you cut it up. Hanging the carcass in a cool shed or barn out of the sun is perfect. If the weather turns warm try to find a place with refrigeration in which to hang your beef. If the weather turns cold enough to freeze the meat, you may be able to quarter and hang it in a protected area such as a cellar way or garage. The ideal place to hang meat (besides a walk-in cooler) is a tight yet well-ventilated building on the north side of the house that can be opened at night to cool and closed in the daytime to keep the coolness in.

After a week or ten days of hanging, your beef is ready to cut, unless you have been forced by warm weather to do so sooner. If meat was cooled rapidly and completely during the first 24 hours after slaughter, it can stand a day or so of up to 50°, but if the meat is not properly cooled it is better to cut it up immediately than to risk spoiling.

First, quarter the split carcass by cutting between the last two ribs of each half. You'll need help to do this, but even

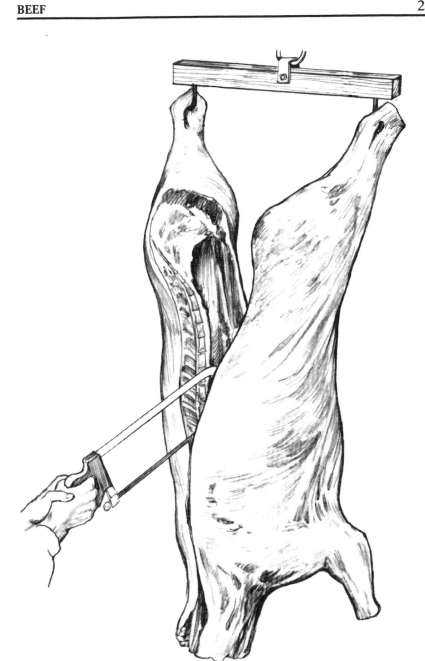

Figure 2.10: Split the carcass by sawing down the length of the backbone, leaving the two equal halves joined by the last few inches of neck tissue.

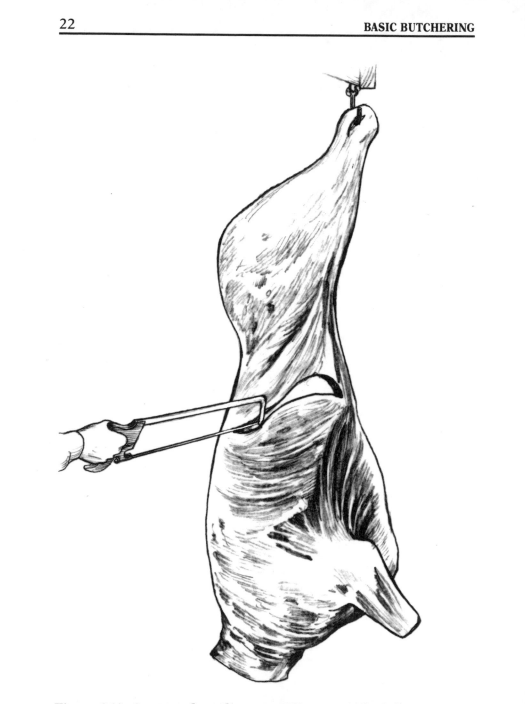

Figure 2.11: Quarter the split carcass by cutting between the last two ribs on each half of the carcass. Leave some flank to hold the quarter together while you saw through the backbone.

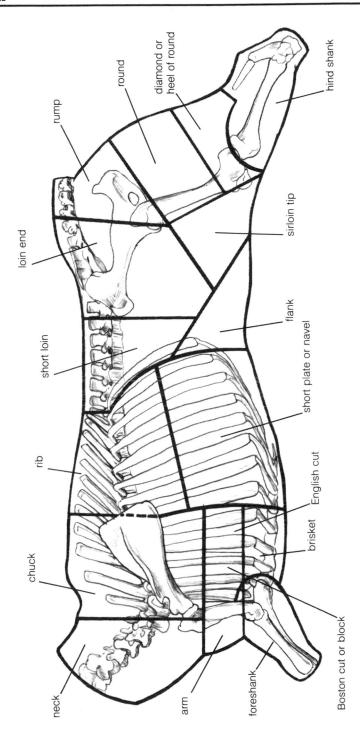

Carcass diagram.

round

diamond or
heel of round

hind shank

rump

sirloin tip

loin end

flank

short loin

short plate or navel

rib

English cut

brisket

chuck

Boston cut or block

neck

arm

foreshank

so, leave some flank to hold the quarters together while you saw through the backbone (figure 2.11).

A good solid table with a wooden top will be needed on which to cut the meat. If you construct your own you can make it a comfortable height to work on. Three-quarter-inch plywood makes a good work surface since it is easily cleaned and is not apt to splinter. Butcher blocks are fine too, and big enough for most meat; but for beef you will need a table at least 3-by-4 feet or larger.

You are now ready to cut up the first forequarter of your beef, producing ribs, plates, steaks, and other cuts. Each bold letter refers to a corresponding illustration which demonstrates that particular procedure.

2.A. The forequarter should be laid on the table inside up. Make a cut with a knife as far as you can between the fifth and sixth ribs, counting back from the front.

2.B. Turn the quarter over and with your saw continue the cut you started in 2.A, dividing the shoulder blade, backbone and breastbone. You now have two pieces, the rearmost being the rib.

2.C. With a knife, and finishing with the saw, cut through the rib about ⅔ of the way between the top and bottom. You now have rib (1) and short plate or navel (2).

2.D. You may now make rib steaks (1) or Standing Rib Roasts (2), or you can bone out the piece into a Rolled Rib Roast (3). In order to make the rolled rib, besides removing ribs and backbone (4), you must remove the yellow back strap and the cartilage that makes up the top of the shoulder blade.

2.E. Use the plate to cut into 2-inch strips for shortribs. Bone the rest for stew or use as soup bone.

2.F. Cutting parallel to the backbone, separate the shank (1) and brisket (2) from what remains of the forequarter. Bone the brisket as pot roast or, better yet, corn it (see page 148). Divide the rest of the forequarter into the arm (3) and block (4).

2.G. Separate the arm (1) from the block (2) by cutting behind and parallel to the bone in the arm. Bone the arm and, after removing extra fat, roll for pot roast.

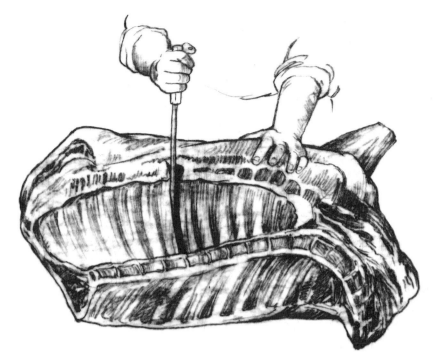

Figure 2.A: Lay the forequarter on the table with the inside facing up. Counting back from the front, cut with a knife between the fifth and sixth ribs.

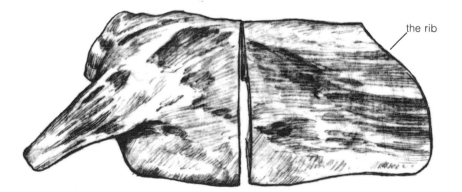

the rib

Figure 2.B: Turn the quarter over, and with a saw continue the cut you started, which will divide the shoulder blade, backbone, and breastbone.

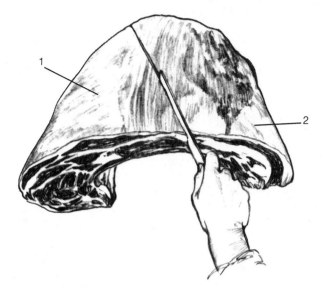

Figure 2.C: First with a knife, and then with a saw, cut through
the rib about ⅔ of the way between the top and the bottom. You
now have the rib (1) and the short plate or navel (2).

Figure 2.D: From the rib you can now make rib steaks (1), Standing
Rib Roasts (2), or you can bone out that piece to make a Rolled
Rib Roast (3). Before you can make a rolled rib, you must remove
the ribs, backbone (4), yellow back strap, and cartilage that makes
up the top of the shoulder blade.

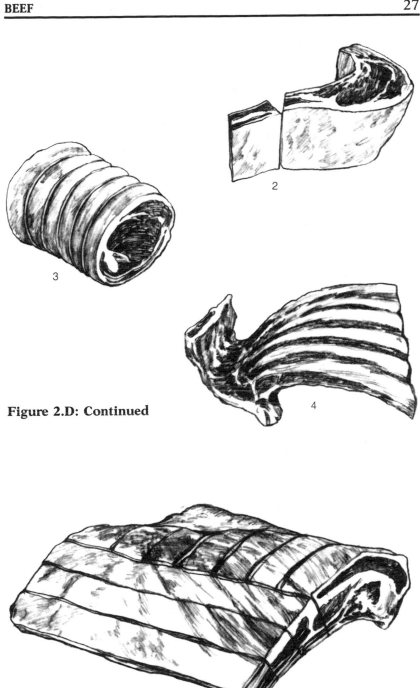

Figure 2.D: Continued

Figure 2.E: Cut the plate into 2-inch strips to make shortribs.

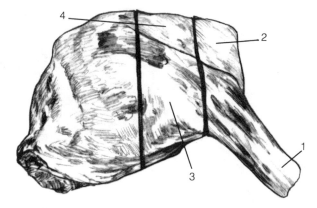

Figure 2.F: From the other half of the forequarter, make two cuts parallel to the backbone. You will work first with the shank (1), the brisket (2), the arm (3), and the block (4).

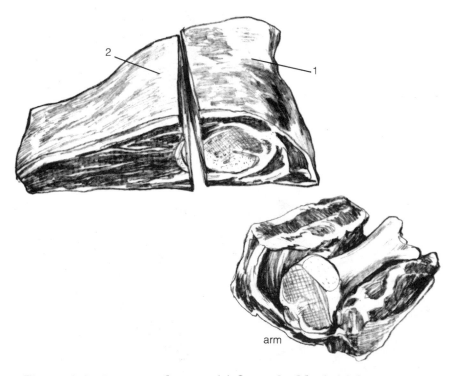

Figure 2.G: Separate the arm (1) from the block (2) by cutting behind and parallel to the bone in the arm. Bone the arm, remove excess fat, and roll for pot roast.

2.H. Separate the block (1) from the English Cut (2) leaving two ribs on the English Cut. With rib removed these cuts are also referred to as Boston Cut or Cross Rib.

2.I. Working on the head end of the forequarter, separate the neck (1) from the shoulder just ahead of the shoulder joint, and the chuck blade (3) from the shoulder (2) by a cut just behind the joint. Use neck and shoulder as pot roast, stew meat, or ground beef. Cut the chuck blade into steaks or bone out and roll into pot roasts.

While you have been cutting the forequarter a helper can have been wrapping, labeling, and putting meat in the freezer. The sooner that is done the better. Stew meat should be trimmed of most fat and other sinewy or bony tissue. Meat for hamburger should be ground, wrapped, and frozen as soon as possible, too. If you do not have help enough to do this, cut the forequarter only into major cuts, keeping them cool until you have time to cut and wrap them into smaller serving sizes.

To start cutting up the hindquarter, lay it inside up and trim out the kidney fat and kidney. Remove arteries, veins, and fat from around the kidney, but if you're going to freeze it leave the membrane covering each lobe in place so the tissue will not dry out (freezer burn) so badly.

2.J. Make one long cut to remove the flank (1), varying the distance from the eye of the loin (2) to govern the size of the tail on the Sirloin steaks.

2.K. The Flank Steak is an oval-shaped muscle that is found in the flank and is too often missed or wasted. Although the rest of the flank is used as stew or hamburger, this cut, sliced on the diagonal, makes delicious London Broil.

2.L. Dividing the remainder of the hindquarter, make a cut parallel to, and about 1 inch from, the aitchbone (pelvis) (1). Separate the rump (2) from the round (3).

2.M. Make 2 cuts parallel to the long bone (femur) (1) of the round from the stifle or knee joint to the cut in 2.L. This will make it possible to remove the tip of the round as shown next.

2.N. By cutting through the stifle joint you may pull the tip of the round free.

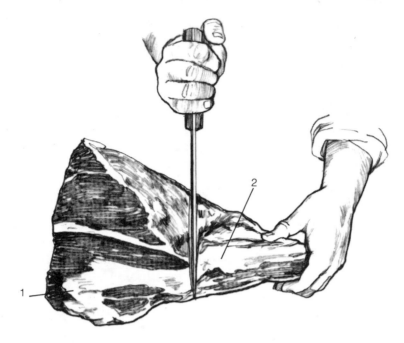

Figure 2.H: Separate the block (1) from the English Cut (2), leaving two ribs on the English Cut.

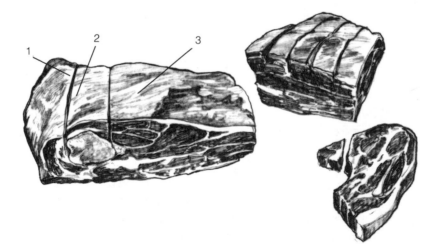

Figure 2.I: Working on the head end of the forequarter, separate the neck (1) from the shoulder just ahead of the should joint, and separate the chuck blade (3) from the shoulder (2) by a cut just behind the joint. Cut the chuck blade into steaks or bone and roll into pot roasts.

Figure 2.J: Working now on the hindquarter, make one long cut to remove the flank (1). Vary the distance from the eye of the loin (2) to govern the size of the tail on the Sirloin Steaks.

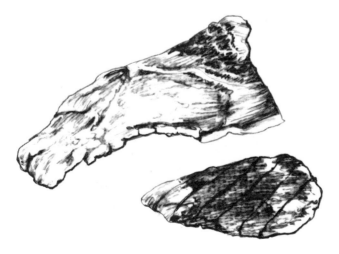

Figure 2.K: The Flank Steak is an oval-shaped muscle found in the flank that is often missed or wasted. This cut makes delicious London Broil.

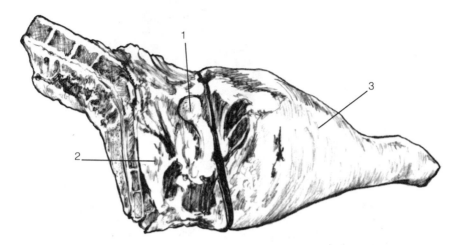

Figure 2.L: On the remainder of the hindquarter, make a cut parallel to, and about 1 inch from, the aitchbone (pelvis) (1), and separate the rump (2) from the round (3).

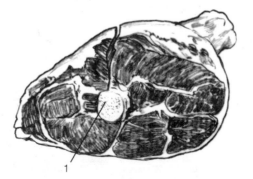

Figure 2.M: Make two cuts to the long bone (1) of the round, from the stifle or knee joint to the cut in figure 2.L.

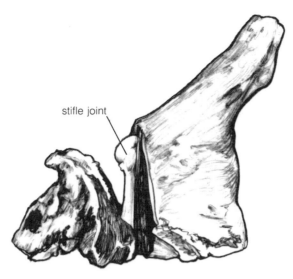

stifle joint

Figure 2.N: By cutting through the stifle joint you may pull the tip of the round free.

2.O. Make a cut as shown parallel to the end of the round about 1 inch above the stifle. Cut through the femur bone with your saw.

2.P. Remove the femur from the round and then divide it into bottom (1) and top (2) round, or outer and inner. You will find a natural division between the single muscle of the top round and the bottom round, which appears to be two muscles.

2.Q. You may make steaks or roasts out of bottom round (1), top round (2), and tip (3). Top round is of highest quality, bottom excellent but not quite as tender, and tip is broiling quality only in better-quality animals.

2.R. Cut the gambrel tendon loose and, by following the shank bone with your knife, remove the heel of the round for pot roast. Trim the rest of the shank for stew or ground beef. The shank bone makes a good soup bone.

2.S. Make a cut with knife and saw as shown about 1 inch in front of the aitchbone (1). This will separate the rump from the loin.

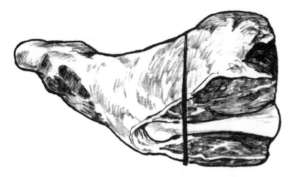

Figure 2.O: Make a cut as shown parallel to the end of the round about 1 inch above the stifle. Cut through the femur bone with a saw.

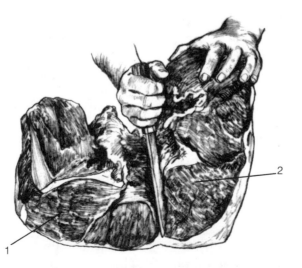

Figure 2.P: Remove the femur from the round and then divide it into bottom round (1) and top round (2).

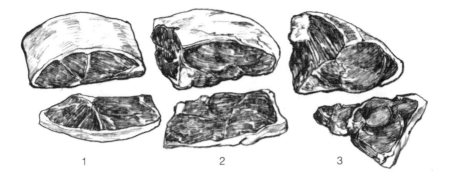

Figure 2.Q: You may make steaks or roasts out of bottom round (1), top round (2), or tip (3).

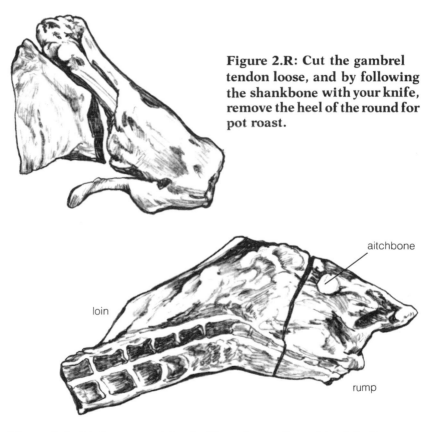

Figure 2.R: Cut the gambrel tendon loose, and by following the shankbone with your knife, remove the heel of the round for pot roast.

aitchbone

loin

rump

Figure 2.S: Make a cut with a knife and saw about 1 inch in front of the aitchbone. This will separate the rump from the loin.

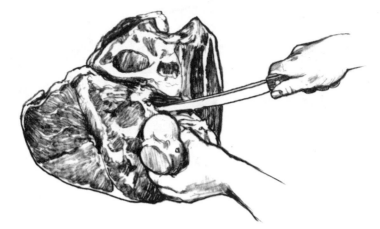

Figure 2.T: Bone and roll the rump, following the aitchbone with your knife. Remove the ball of the ball-and-socket hip joint and all bone as you come to it.

Figure 2.U: Divide the loin into Club (1), T-bone (2), Porterhouse (3), and Sirloin Steaks (4).

2.T. Bone and roll the rump, following the aitchbone with your knife. Remove the ball of the ball-and-socket hip joint and all bone as you come to it. Use metal skewers to hold the rump roast together.

2.U. The loin is divided into Club (1), T-bone (2), Porterhouse (3), and Sirloin steaks (4).

The first beef you cut will seem difficult, but as you cut you will learn. The foregoing describes one way to do the job; you will find little short cuts and different methods that better suit your skill and the way you ultimately consume the beef. Any really bad mistakes in cutting can always be made into stew or hamburger, so nothing is wasted.

Ground beef, or hamburger, should have 4 parts of lean meat to 1 part fat. Remember that when you trim for grinding. Fat does not keep well in the freezer, and it is better to throw some fat out or buy some lean lower-quality beef to grind with the fat than to try to freeze too-fat ground beef. Further, if there is any question as to the quality of a piece of meat (dirty, bloodshot, dry, or foul smelling) don't grind it, because it will spoil the entire batch.

3·HOGS

HOG KILLING IS AS MUCH an American tradition as Thanksgiving or Fourth of July. In fact, in the South it used to be done on or near Thanksgiving Day, and in some rural areas is still done at that time.

It is almost impossible for an individual to slaughter a hog or hogs without help. Neighbors and family members have always helped each other at hog-butchering time, making it an almost festive occasion, like a barn raising or husking bee. As a boy, I recall that hog-killing day was one day when I was allowed to skip school, since the more help available, the better the job went.

Getting Ready

Quite a lot of paraphernalia and preparation are required for hog butchering, so once you are set up and have help available it isn't much more difficult to do several hogs than it is to do one. Even if there is a slaughterhouse in your area that will custom slaughter, you still have the most difficult job of getting the live pigs out of the pen and to the slaughterhouse. If you have help enough to do that, why not kill and stick the pig right in the pen, then drag it out, and scald and scrape it. The rest isn't any more difficult than butchering a lamb or veal, so why not try?

To butcher hogs you will need a large tank to hold water.

Few old cast-iron potash kettles are still available, but a 55-gallon barrel, wooden or metal, can be used, as well as a 95-gallon stock tank. If you are going to do a lot of slaughtering the latter is preferred, and should be set in a concrete block or brick foundation over a firebox.

Next to the water tank you need a wooden plank platform 4½ feet wide and 6 feet long with the 4½-foot end toward, and at the height of, the top of the tank. A length of rollers scrounged from a nearby mill or junk yard, will make it easier to get the dead hog up onto the platform. A block-and-tackle or come-along to lift the hog up onto a hanging pole is almost a necessity, but if you butcher your hogs at no more than 180 pounds and have lots of willing help you can get along without it. A little imagination and ingenuity can add other labor-saving schemes. For example, a door of the hog pen should be small so that when you take the dead hog out of the pen the others can't escape. An electric outlet or extension cord is convenient for using a power meat saw, and even simple things like a nail on which to hang the meat saw and a handy water hose to wash things down make the job easier. A rubber apron and hip boots for the people scraping the hog make for more comfortable working conditions.

The only tools you *must* have are: bell scrapers; a curved or double-edged knife for sticking; a meat saw or clean, sharp carpenter's saw; a small, sharp butcher knife; a hog hook or a two-handed hay hook; and either a singletree or gambrel stick.

The hog to be butchered should be about 180 pounds live weight. The ideal day for butchering is clear and still, below 40° but above 20°. Obviously you can't always have such a day when you're ready, but try to aim for one. Some authorities say you should starve pigs out for 24 hours before slaughter, but allow them all the water they want. That, too, is easier said than done, since hungry hogs may tear the pen down, and even if you have no affection for hogs you feel sorry for them. To be practical, feed in the early afternoon the day before butchering but, of course, skip the morning feeding.

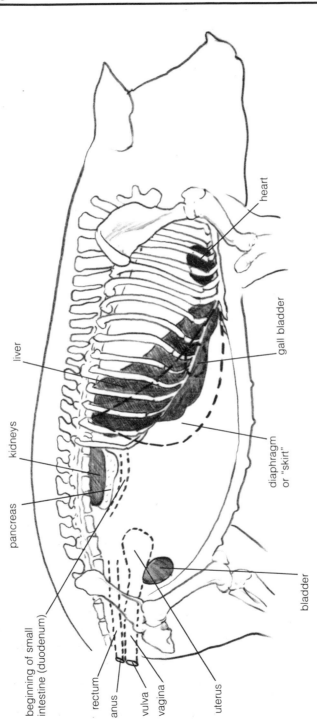

heart

gall bladder

liver

diaphragm or "skirt"

kidneys

pancreas

bladder

beginning of small intestine (duodenum)

rectum

anus

vulva

vagina

uterus

Anatomical diagram.

Killing and Scraping

Hogs may be stunned, but it is easier to shoot them with a .22. When the water in the tank is hot (145° to 165° according to a thermometer floating on a board), and all is ready, coax a hog to a clean end of the pen with a little dry feed and, as it eats, shoot it just off center from where an X between eyes and ears intersects **(figure 3.1)**. The hog should drop and, as it does, you hand the gun to a person behind you who is holding the sticking knife. You and a third person should then roll the hog on its back. That person grabs its front feet and spreads the legs, holding the hog up on its back with a knee against its ribs or by straddling it. You grab the hog's lower jaw, holding its head down. The number-two person then hands you the knife, and you make a cut 2½ inches long lengthwise of the hog's body just ahead of the breastbone. Now push the knife, cutting edge down, until you hit the breastbone, and go under the breastbone at about a 45° angle. Then, using the breastbone as a fulcrum, push the handle toward the hog's rear. That will push the blade forward toward the animal's head and down toward the backbone **(figure 3.2)**. Again using the breastbone as a fulcrum, cut sideways toward the first rib on each side and then withdraw the knife. Blood should be literally pouring out, but although you have cut the carotid arteries you will not see blood spurt as it does in other animals when the throat is cut.

An alternative method of restraint is to shackle one leg of the hog with a chain or rope after it is shot and raise it off the ground. Sticking is then done in the same manner.

As soon as you are sure the animal is dead (pour a little water in its ear if you are not sure), put a hook in its mouth, into the lower jaw under the tongue, and drag it out of the pen (through the small door mentioned earlier) and up onto the wooden platform. Although immersion in 140° water for about 4 to 6 minutes will loosen hair, on a cold day you'll do well to start with water at 165°. Lower the hog's rear end into the tub **(figure 3.3)**. Wood ashes, lime, rosin, and pine tar are agents that can be added to the water to make the

Figure 3.1: Shoot (or stun) the hog just off center of the place where the lines shown intersect.

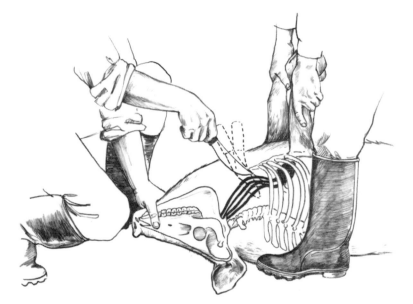

Figure 3.2: When the hog drops hand the gun to a second person, and then a third person should roll the hog onto its back and spread its front legs apart. You then grab the hog's jaw and hold its head down. The number-two person hands you the knife, you make a 2½" vertical cut lengthwise of the hog's body just ahead of the breastbone. Now push the knife, cutting edge down, until you hit the breastbone, and then go under the breastbone at a 45° angle. Using the breastbone as a fulcrum, push the handle toward the hog's rear — this action will push the blade in the direction of the animal's head and down toward the backbone. This is a difficult procedure — be sure to familiarize yourself with the text on page 42 before you attempt it.

hair stick to the scrapers better, but all in all, the water temperature is the most important factor in loosening hair.

With the dipping water at 165°, keep the hog moving up and down so it doesn't cook. As hair begins to loosen, usually in 2 to 4 minutes, slide the hog out and rub hair from around its feet with your hands. Then while two people scrape with bell scrapers **(figure 3.4)**, tilted and working with the hair as you would shave, take your sharpest knife and "raise the gambrels" by cutting lengthwise just above the hog's dewclaws on the rear leg and exposing the tendons **(figure 3.5)**. At the same time you might use a hook to snap off the horny part of the hoofs and dewclaws, and put a gambrel stick or singletree in the gambrel to raise the hog to its hanging position.

Figure 3.3: Put a hook in the hog's lower jaw under the tongue, and lower the rear end of the animal into hot water. Keep the hog moving up and down so that it won't cook.

Figure 3.4: Scrape the hair off the hide with bell scrapers. Tilt the scraper and work the hair off in the same way that you would shave.

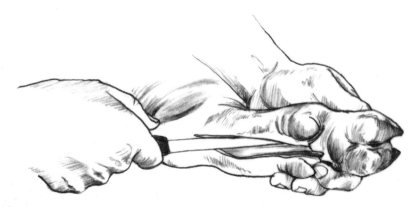

Figure 3.5: Using a sharp knife, cut just above the dewclaws on the rear legs to expose the tendons by which you will "raise the gambrels."

Fire pits, supports, and roller — in preparation of a hog butchering.

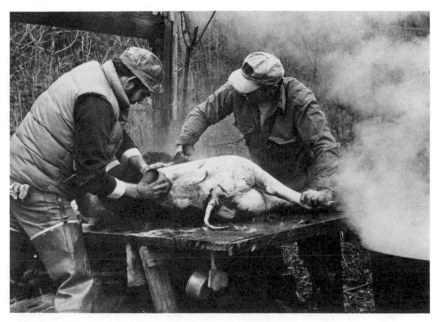

Using bell scrapers to take the hair off the hog carcass.

By now the people scraping the rear end should be finished, and you are ready to dip the front end. Of course, if you have a big enough tank, such as a 95-gallon stock tank, you might have dipped the whole hog at once.

After the front end is dipped and cleaned, rinse the hog with warm water and rescrape with the bell scrapers held flat. You can now either "shave" with a sharp knife or singe with a blow torch to get off the few bristles you've missed.

While the hog is still on the platform insert the knife, blade up, into the sticking place and cut the skin on the midline all the way to the rear end of the breastbone. With the meat saw split the breastbone, being careful not to go beyond the chest cavity where you might cut into stomach or intestines **(figure 3.6)**.

Now, in the opposite direction, with a sharp knife find the gullet and windpipe, cut on either side, and remove them as

Figure 3.6: Split the breastbone with a meat saw. Try to keep the saw as parallel to the line of the breastbone as possible to prevent cutting the inner organs, and do not cut beyond the chest cavity.

far forward as you can. Also dissect the tongue. Remove the tongue and put it in a clean enamel pot or hang it to cool.

While the hog is still on the platform, or after hanging, split it between the hams, first with a knife, trying to find the division between the two sides of the pelvic bone. If the hog is older, and you can't split the bone with a knife, use the saw, but be careful not to puncture the bladder. If the hog is a barrow (castrated male), loosen the penis, letting it hang to be removed with the bung. To loosen the bung, cut around the anus from behind. If you are working on a female (gilt), the vulva and the genital tract are included with the bung. Tying a string around the bung at this time to prevent manure spilling is not always necessary, but it's a good idea just in case of accident.

Either now or after you hang the hog by the gambrel stick or singletree, cut the belly wall from the pelvis, forward from inside out, being careful not to puncture intestines or stomach **(figure 3.7)**. With the hog hanging, reach up and grab the bung with your left hand and either by blunt dissection with your right hand, or with your knife, loosen the bung and pull it down with the bladder and penis. Still working the same way, use your left hand to pull the intestines down and out, and use your right hand with or without the knife to tear them loose from their attachments **(figure 3.8)**. If you have not punctured intestines or stomach, keep working the same way, cutting the diaphragm with a semicircular cut parallel to the shape of the body. Loosen the lungs and heart, cutting the aorta (large blood vessel just under the backbone) loose if necessary. Keep pulling with your left hand and loosening with your right until you have removed everything from the bung to the place where you cut the gullet loose to remove the tongue.

If the liver stayed in, loosen, remove, and rinse it. Then, after dissecting out the gall bladder, rinse again and hang it to cool. If the liver came out with the intestines salvage it, and also, if you can find it, the "belly sweetbread," the pale pinkish-orange pancreas.

Wash and rinse the carcass inside and out with warm water. To facilitate cooling remove as much belly fat as possible.

Figure 3.7: Cut the belly wall from the pelvis — forward from the inside out — to the hog's snout. Be careful not to puncture intestines or the stomach.

Figure 3.8: Use your left hand to pull the intestines down and out, and with the knife in your right hand (if you are right-handed), cut the intestines loose from their attachments.

Kidneys may be left in the carcass along with a little fat to keep the tenderloins from drying out.

Splitting and Butchering

The carcass will cool faster if split now. This may be done by hand or with an electric power saw if you have access to one. Simply split the backbone from the tail to, and including, the head and through the snout **(figure 3.9)**. This will expose the brain, which is edible, so that it can be easily removed. To prevent the sides from falling off the gambrel stick when split, either tie the sides on or have help to catch them. A singletree is more practical than a gambrel stick at this point because the hooks at each end prevent the sides from sliding off.

By tradition nothing from the hog is wasted, snout to tail. If you care to, you might have saved the blood as it pumped out, catching it in a pan or pail and, adding a handful of salt to keep it from clotting, to make bloodwurst. The meat from the head and snout can be used for headcheese. You should save the heart, tongue, liver, and kidneys. Small intestines are saved, rinsed, soaked in salt water, turned inside-out to remove the soft inner lining, and soaked again in clean salt water. Then they are drained and stored, and packed in salt, in a cool place, to be used as sausage casing. The hooves may be boiled to make gelatin and, of course, the pig's feet pickled. The spleen or "meltz" is sometimes confused with the liver and is usually thrown out, but even the tail is boiled and eaten. Children used to like to roast it. If you really want to be a traditionalist, blow up the bladder with a straw, tie it off, and let the kids have it for a "colonial football."

Hang the sides of the split hog at 30° to 40° overnight to cool and then start to cut them up. There is nothing to be gained in hanging pork for days as you do other meats.

To cut pork you need only a meat saw or sharp clean carpenter's saw, a narrow-bladed boning knife, a sharp butcher knife, and a steel to keep the knives sharp. A wooden topped table large enough to lay the side of pork on, and at a comfortable height to work at, is needed also.

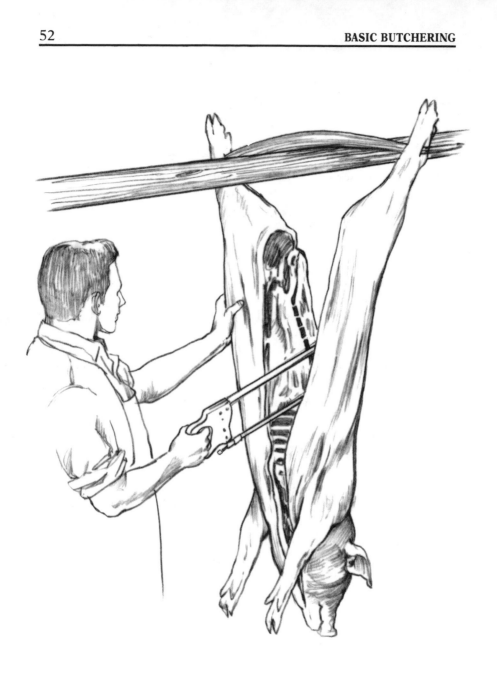

Figure 3.9: Split the backbone with a meat saw from the tail to the head and snout.

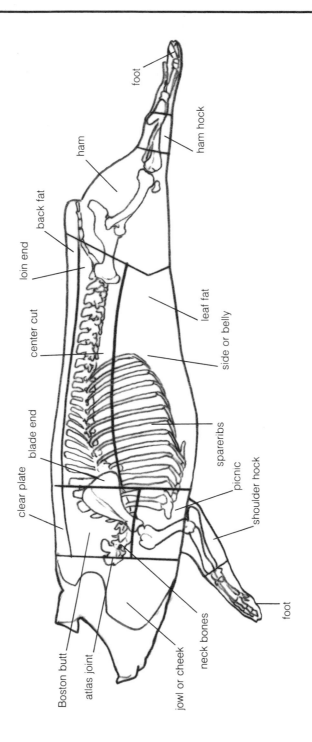

foot

ham hock

ham

back fat

loin end

leaf fat

center cut

side or belly

blade end

clear plate

spareribs

picnic

shoulder hock

foot

Boston butt

atlas joint

jowl or cheek

neck bones

Carcass diagram.

3.A. Remove the head by cutting and sawing through the first joint behind the skull. The jowl is cut from the head and can be put into headcheese or sausage, or smoked (see pages 144–53). The shoulder is removed by cutting in a line parallel to the head-removal line and through the third rib, counting from the front.

3.B. Separate the shoulder into Picnic (1) and Boston Butt (2) as shown.

3.C. Remove the heavy fat covering from the Boston Butt, leaving about ¼-inch of fat covering the butt. Cut the butt into slices, or bone and roll it.

3.D. Separate the breast (1) from the Picnic (2) using the breast meat for sausage.

3.E. Remove the foot (1) as shown and the shoulder hock (2) (actually the elbow, not the heel). That leaves the Picnic squared off; it can be boned and prepared with pocket or rolled.

3.F. If you would rather leave the shoulder whole to smoke, do not separate the Boston Butt from the Picnic, and trim as shown.

3.G. At the rear of the hog, before separating the ham from the loin, study the carcass and find the backbone (1) and the aitchbone (2). Now, depending on how large you want the ham, make a cut perpendicular to the shank that will give you the size ham you want. You may angle the cut at (3) to give a longer side for more bacon.

3.H. Separate the side (1) from the loin (2) as shown.

3.I. Start to cut the fat off the loin with a full-length steady pull of the knife; turn the loin over and do the same to complete the job. Leave about ¼-inch of fat on the loin. The loin may be cut into chops, roasts, or both.

3.J. Trim the spareribs from the side.

3.K. Cut a strip off the side to remove the nipples and square the cut. The side is cured and smoked as bacon.

3.L. Trim the ham, removing excess back fat, backbone and tail. Leaner trimmings may be used for sausage. Remove ham hock (1) and foot (2) as shown.

3.M. Use the ham whole, fresh, or smoked, or cut into roasts or steaks, or ham may be boned and rolled.

Excess fat may be used for lard.

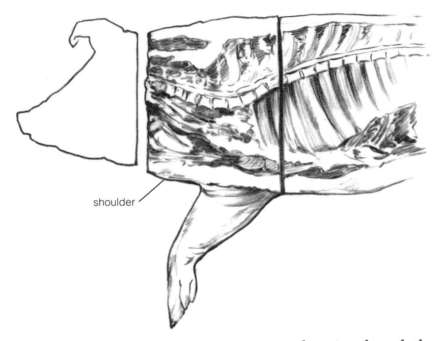

shoulder

Figure 3.A: Remove the head by cutting and sawing through the first joint behind the skull. Remove the shoulder by cutting in a line parallel to the head-removal line and through the third rib from the front.

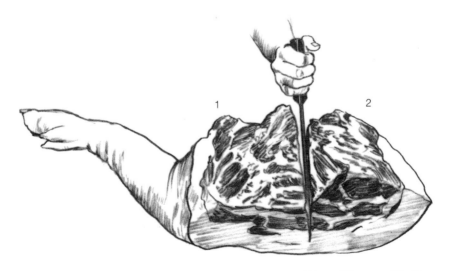

1

2

Figure 3.B: Separate the shoulder into Picnic (1) and Boston Butt (2), as shown.

Figure 3.C: Remove most of the heavy fat that covers the Boston Butt, but leave about ¼-inch of fat to cover it.

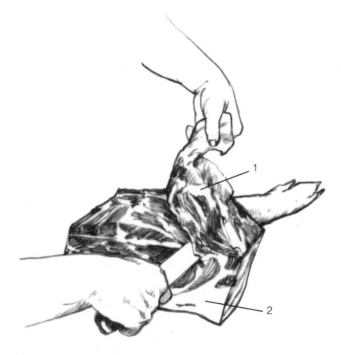

Figure 3.D: Separate the breast (1) from the Picnic (2).

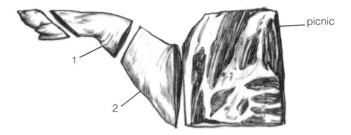

Figure 3.E: Remove the foot (1) and the shoulder hock (2), as shown. That leaves the Picnic squared off.

Figure 3.F: An alternative would be to leave the shoulder whole for smoking. If you choose to leave the shoulder whole, do *not* separate the Boston Butt from the Picnic (figure 3.B.), but trim the shoulder as shown.

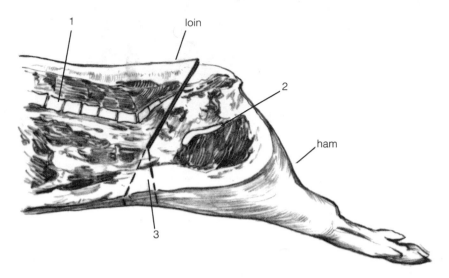

Figure 3.G: Working now on the rear of the hog, you will want to separate the ham from the loin. First, find the backbone (1) and the aitchbone (2). You may angle the cut at (3) to give you the size ham, and the amount of bacon, you want.

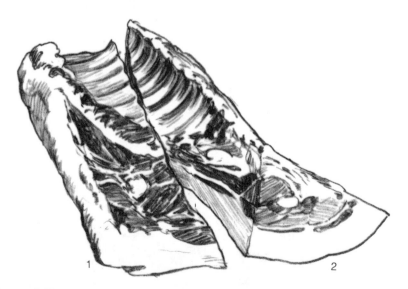

Figure 3.H: Separate the side (1) from the loin (2).

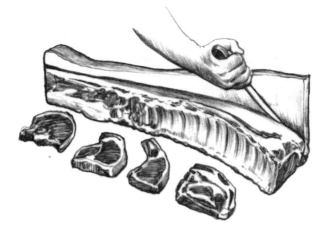

Figure 3.I: Cut the fat off the loin with a steady pull of the knife along the length of the loin; turn the loin over and do the same thing to complete the job. Leave about ¼-inch of fat on the loin.

Figure 3.J: Trim the spareribs from the side.

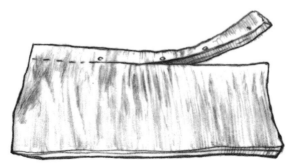

Figure 3.K: Cut a strip off the side to remove the nipples and square it off.

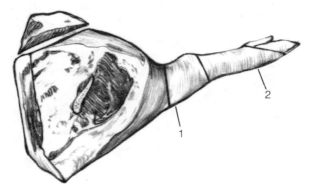

Figure 3.L: Trim the ham, and remove excess back fat, backbone, and tail. Remove the ham hock (1) and foot (2).

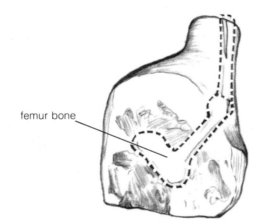

femur bone

Figure 3.M: The whole ham. It may be served fresh, smoked, cut into roasts or steaks, or boned and rolled.

4·VEAL

THE BEST VEAL is that from a calf raised on a nurse cow for about 12 weeks. Under modern farm conditions this is usually impractical and expensive, yet there are times when it is done. The next-best option is a veal calf fed only milk until it is 12 weeks old. For someone who has a family cow that is a very practical option, since when the cow is fresh she gives more milk than the family consumes. By the time you have fatted two calves, or started a beef steer or replacement heifer and fatted one calf, the cow's production has dropped. Never try to milk a cow half way and let the calf nurse the rest or vice versa. You'll end up with mastitis, a dry cow, a sick calf, or all three. On a dairy farm one can fat a calf on excess milk such as pipeline rinsings or other unsalable milk. And there are, of course, milk replacers meant just for fattening veal, and if used properly they do produce fine veal.

In no case should you try to make veal out of just any 12-week-old calf raised as you would a dairy calf on hay, grain, and milk or milk replacer, or out of a calf much over 12 weeks of age. These animals are apt to be stringy and tough.

Bob or newborn calves sent to slaughter end up as so-called baby veal, but I have no desire to eat that (although I probably have done so unknowingly in restaurants), and would not recommend that you butcher one. Of course, if you have a calf well fatted on milk at 4 weeks of age and you lose your milk supply, it would be better to butcher it at that time rather than feed it a conventional diet of hay, grain, and regular milk replacer, ending up with a rangy, stringy 12-week-old animal. Calves over 12 weeks old that have been nursing their mothers are sometimes used as veal. In fact, many used to come into slaughterhouses in New York City and were jokingly called "swamp veal" because they usually came from Louisiana or Mississippi. These animals, of Brahmin ancestry, could run across a slippery killing floor like a deer across a sodded field. I would guess that that veal took a lot of pounding before it was tender enough for choice Wiener schnitzel. Butchering for your own consumption gives you the choice of avoiding such poor-quality meat.

Getting Ready

To butcher veal all you'll really need is one good sharp knife and some rope. You should have string to tie the bung, and you may need an extra knife, a steel and sharpening stone, and a meat- or carpenter's saw. A clean pail and a source of water, preferably from a hose, are also a help, as are soap and paper towels. A clean container in which to put liver, heart, and sweetbreads should be within reach when you need it. For cutting up the meat later you will find the work easier with a small, narrow-bladed boning knife.

As with most animals, starving a veal calf out for 24 hours before slaughter is recommended; but to be practical, with-holding the morning milk feeding on the day of slaughter is all that is usually done.

The most difficult part of butchering veal is killing the animal. After you have fed and cared for a calf for 12 weeks you will find yourself looking for excuses not to butcher it. Therefore, although a veal is easily butchered alone, it's a good idea to have assistance, so someone else can do the actual killing.

Killing and Removing the Organs

To determine the spot to shoot or stun on a calf, draw an imaginary line from each eye to the opposite horn (or where the horns would be). Just to the left or right of where the lines would cross is the place to shoot with a .22 or hit using a small sledgehammer **(figure 4.1)**. The most humane method of killing is the quickest. If you are unsure of stunning and don't feel comfortable using a gun, you can do a humane job of killing by simply hanging the calf by its hind legs and cutting the throat just behind the jaw, literally from ear to ear, with a sharp knife **(figure 4.2)**. Be sure you sever both carotid arteries as well as the jugular veins. In doing so you will, of course, cut the windpipe and gullet. Death will come quickly, and the calf will bleed well when it is hung this way. If you do shoot or stun the animal, be ready to hang it immediately and cut the throat as described.

Veal is left with the hide on until it is ready to be cut up, so only remove the hide from the front legs as far back as the knee joint and remove the leg at the knee, leaving the hide on to keep the exposed surface clean. Then lower the calf, and while it lies on the ground, cut around the bung, loosen it, and tie it off with a string so it will not leak manure. Next skin the rear legs from the dewclaws back to just above the hock (gambrel) joint and remove the leg between hock and cannon bone below that joint **(figure 4.3)**.

Figure 4.1: Shoot or stun the calf just to the left or right of the place where the lines shown intersect.

Figure 4.2: Another way to kill the calf is to hang it by its hind legs and cut the throat with a sharp knife, just behind the jaw, from ear to ear.

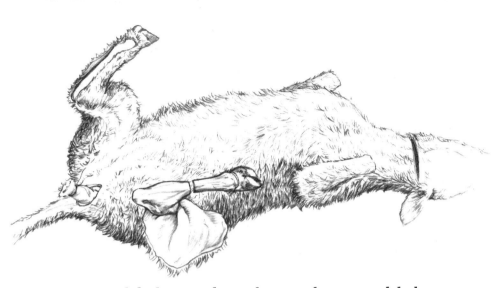

Figure 4.3: While the animal is on the ground, cut around the bung, loosen it, and tie it off with a string. Then skin the rear legs from the dewclaws to just above the hock joint, and remove the leg between the hock and the cannon bone. Cut off the front legs at the same place as well.

Now raise the carcass with a gambrel stick or short single-tree hooked into the gambrels **(figure 4.4)**. Make a cut through the skin and, cutting from the inside out so as to avoid getting hair all over the meat, open the hide as far up as the testicles (or udder) and as far down as the cut in the throat. Next loosen the skin back 2½ inches on each side of this cut. Now carefully open the belly wall on the midline just ahead of the pelvis **(figure 4.5)**. Once air rushes in the insides will drop down. Then you can more easily cut from the inside out, using your fingers as a guide, splitting the belly on the midline as far down as the breastbone. The breastbone can be split on the midline with a sharp knife also, as far down or forward as it extends.

Reach into the pelvis from the front and hook a finger or fingers around the bung. Using blunt dissection and your knife, pull it down and out. (An alternative is to split the pelvis, but this exposes more meat to drying when you must remove skin over the pelvic area.) Holding the bung in your

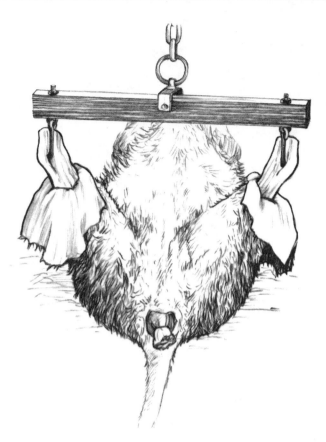

Figure 4.4: Hook the hocks of the rear legs into the gambrel space from a singletree, and then raise the carcass.

left hand (if you are right-handed) use your right hand to peel and cut the intestines loose from their attachments, pulling the whole mess out, including the stomach (but not the liver or kidneys) **(figure 4.6)**. Avoid puncturing the digestive tract if possible. Either cut the whole mass loose at the gullet or, if it is not too full and heavy, let it hang while you look for and remove the belly sweetbread (the pinkish-orange pancreas) and pull out the liver. After removing the gall bladder from the liver, rinse the liver with warm water and hang it in a cool, clean place with the belly sweets.

Now cut around the diaphragm, as close to the ribs as possible, pull out the lungs and heart, dissect them loose

Figure 4.5: Open the hide with a careful cut from the testicles or udder to the throat. Loosen the skin back 2½ inches on each side of the cut. Now open the belly wall on the midline just ahead of the pelvis to the breastbone.

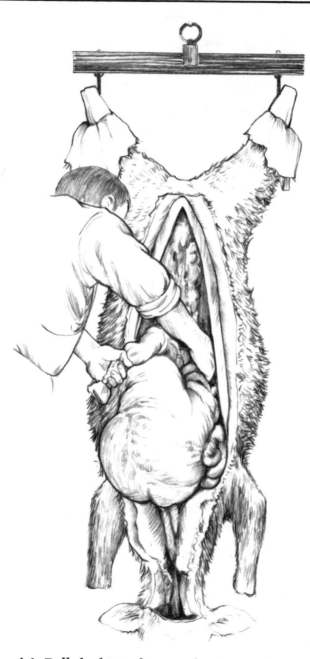

Figure 4.6: Pull the bung down and out into the open cavity. Holding the bung in your left hand (if you are right-handed), use your right hand to peel and cut the intestines loose from their attachments, and pull everything out, including the stomach.

with your knife and, following the windpipe and gullet, cut that out as far down as where you cut the throat. Be very careful as you reach the area ahead of the heart and along the windpipe to find the true sweetbread — the thymus **(figure 4.7)**. If possible leave it there until you have removed the lungs and heart. If it does come out with the lungs and heart dissect it loose to save.

Discard everything else you've pulled out except the heart. Cut through the membranes around the heart and cut it loose from the large blood vessels. Open it and take the clotted blood out. Rinse the heart and put it with the liver and belly sweetbread. Don't forget to go back and look for the sweetbread where it lies in the throat along where you pulled out the windpipe. Dissect it loose, rinse, and put with the rest of the sundries.

You have now removed everything from the carcass except the kidneys, tongue, and brain. The tongue is easily removed by opening the skin under the jaw, following the windpipe down on either side with a knife, and dissecting it out **(figure**

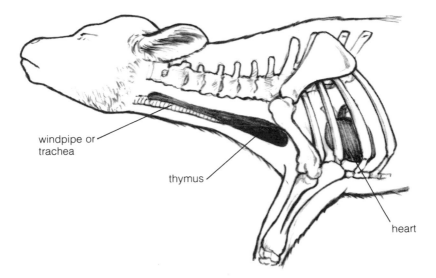

windpipe or trachea

thymus

heart

Figure 4.7: You can locate the true sweetbread (thymus) along the windpipe, and in the area ahead of the heart. Dissect it loose and save it.

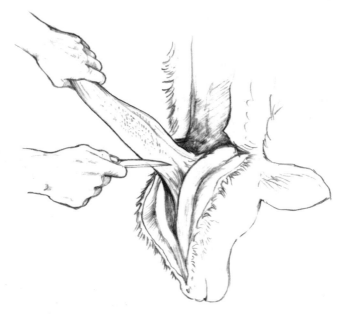

Figure 4.8: The tongue is removed by cutting open the skin under the jaw. Next follow the windpipe down, pull the tongue out, and dissect it loose.

4.8). To remove the brain you will have to split the head, pull the brain out, rinse it, and cool it as quickly as possible. The kidney on each side is usually left until the animal is cut up so the fat surrounding it will protect the tenderloin.

Butchering

Rinse the carcass inside with warm water, put two sticks in to spread it open, and hang it to cool **(figure 4.9)**. The carcass cannot be easily cut up until thoroughly cooled, which takes at least overnight at 32° to 36°. Hanging a veal with the hide off will dry it too much. In recent years it has been said that it does not pay to age veal. However, if you have a cool enough place to hang it, I feel it does improve the meat to hang it for up to a week with the hide on.

If the temperature is above 40° and you don't have a place to hang your veal, by all means skin it and get it cut up as soon as possible.

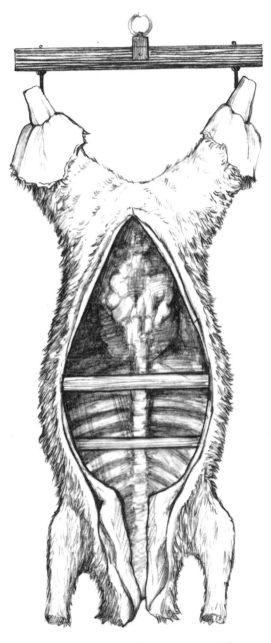

Figure 4.9: Rinse the inside of the carcass with warm water, and put two sticks inside the cavity to spread it open. Hang it and cool it overnight at a temperature between 32° and 36°. *Do not* skin the carcass until you are ready to butcher the meat.

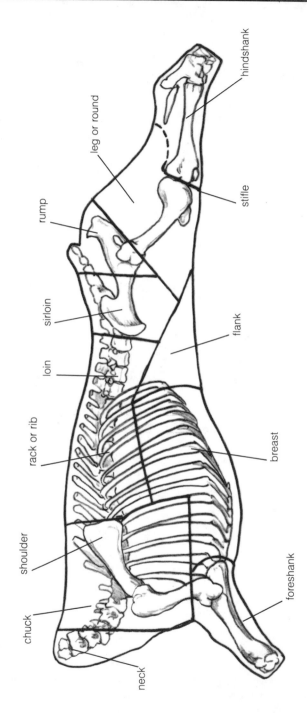

hindshank

leg or round

stifle

rump

sirloin

flank

loin

rack or rib

breast

shoulder

foreshank

chuck

neck

Carcass diagram.

You can see from the diagram the major cuts of veal. Of course, you can vary them to suit your own needs. Split the carcass down the backbone with a saw. Divide it just behind the last rib into forequarter and hindquarter.

4.A. Shown in the drawings are the cuts you will get from the forequarter: shoulder, rib, rib chops, foreshank, and breast. Remove the foreshank, bone it, and cut it into stew. Follow the cut through the ribs at the level at which you cut the shank back to behind the fifth rib. There go up with your cut until you reach the level of half a rib, and cut back to the last rib, removing the breast. The breast can be boned

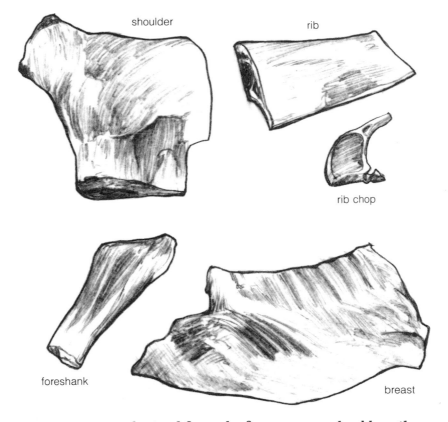

Figure 4.A: Cuts obtained from the forequarters: shoulder, rib chop, foreshank, and breast.

and rolled, or stuffed, or made into stew. Remove the rack, or rib, by continuing the cut behind the fifth rib up through the backbone. The rib may be made into chops or used as a roast. Remove the neck portions of the shoulder to make into stew. The shoulder may be used as is or boned and rolled.

4.B. Cuts from the hindquarter are as shown: leg, rump, loin, cutlet, and chops. Remove the kidney from the fat in the rear quarter and save. Remove the flank at the angle shown in figure A and, if it is not too dried out, use it for stew or grinding. A cut just ahead of the hipbone will remove the loin; cut it into chops or leave whole to roast. Remove

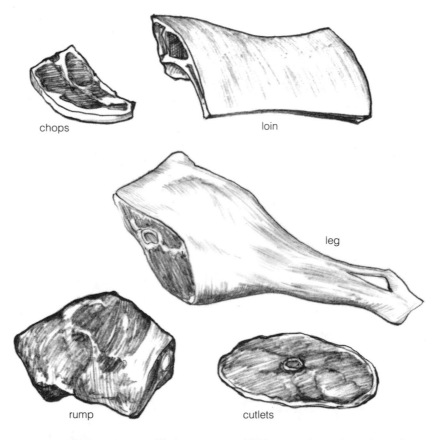

Figure 4.B: Hindquarter cuts: from the leg you obtain rump and cutlets, and from the loin you cut chops.

the leg (or round) from the rump by a cut just below the aitchbone. Make cutlets from the leg until you reach the stifle, then bone that (hindshank) for stew or grinding. The rump may be boned and rolled, but it's better to cut chops from its sirloin end and make the rest into small thin scallops for scallopini veal. In fact, in good veal, much of what is said to be stewing meat, such as neck, shoulder, and shank, can be made into scallops.

Remember that veal dries out and gets freezer burn quicker than some other meats. Double wrap it for freezing.

5·LAMB

THE BEST LAMB for home slaughter is true spring lamb born in March or April and butchered in November. Animals 12 to 14 months old are still lamb; up to 24 months of age they are yearling mutton; and those over 2 years old are mutton. Mutton served in the better chop houses in London certainly tastes different from the mutton of old ewes butchered on the family farm of my youth. I would guess the English mutton is not only younger but better fattened.

Baby lamb or Easter lamb, as used by people of Italian and Greek ancestry, is butchered when 1 to 3 months of age. Butchering of those lambs is often done at home by methods similar to those we are describing here. The carcasses of the lambs are often prepared whole so further cutting is not done.

Getting Ready

Lambs to be slaughtered should be kept in a clean, dry pen overnight without feed, but they should have adequate water. Don't attempt to kill lambs for a day or two after a rain when their wool is wet, heavy, and dirty. A cool day, 25° to 40° is perfect unless you have cooling facilities.

To slaughter you will need one sharp butchering knife, a small skinning knife and, of course, a steel to keep them sharp. If you are butchering only one or two lambs you can work outdoors under a tree with an overhanging limb 8 to 10 feet off the ground. The ground may be covered with straw

77

to keep things clean. If you are working indoors, you should have a solid beam to hang the lamb on, and preferably a concrete floor that can be swept and washed clean. You will need some light ropes to hang lambs, and you could use a small gambrel stick for each lamb. Pails and water should be handy. Cleanliness is important in all butchering, but with sheep it is imperative. You must not touch meat with hands contaminated by manure, dirt and/or wool. Contact with wool gives meat a bad flavor. To "punch" or "fist" off the hide your hands should be wet.

You will need a .22 rifle or pistol if you wish to shoot the lamb; stunning is difficult because of the shape of the animal's skull (**figure 5.1**). However, if you are going to slaughter many animals, a sawbuck rack large enough to hold a lamb placed on its back with its head hanging off on the end is a convenience that will enable you to practically guillotine the lamb, actually beheading it, thus eliminating pain almost as quickly as shooting and avoiding the danger of using firearms in situations where accidents can happen (**figure 5.2**). Of course, you can also shoot the lamb before placing it on the sawbuck, thus having the best of both systems.

To cut the throat after shooting or stunning, make sure the lamb is down and still. Stick a boning knife, blade facing the backbone, just behind the jaw and cut out. This will sever the both jugulars, both carotids, the gullet, and the windpipe. If it doesn't get the carotid arteries, go back with a second cut closer to the bone.

On the sawbuck either strap the lamb down or have someone hold it. Grab the lamb's muzzle, bend the head back just a bit and, with one clean stroke of a sharp butcher knife, cutting down toward the backbone, sever the same structures as above. Twist the head a bit and with the knife disjoint the head from the body where the backbone joins the skull. It may look crude, but the method brings sudden death and allows the heart to keep beating, pumping blood out.

Once the sheep is bled out, regardless of method of slaughter, skin out the front legs, removing the leg at the knee, or better yet at the "break" joint between the knee and ankle if the animal is still young enough to be lamb (distal end of cannon bone or metacarpus). Cut from inside out down the

Figure 5.1: Shoot or stun the lamb as close as possible to the place where the lines shown intersect.

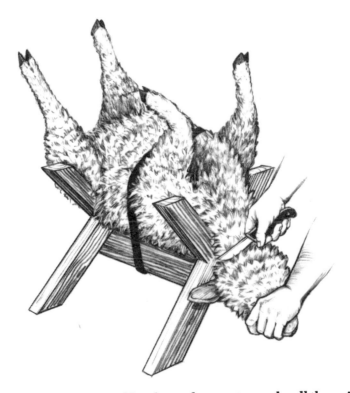

Figure 5.2: After the animal has been shot or stunned, roll the animal onto its back (on the ground or strapped onto a sawbuck), and cut the throat just behind the jaw and across from ear to ear. If you have not cut both the jugulars and the carotid arteries, go back with a second cut closer to the backbone.

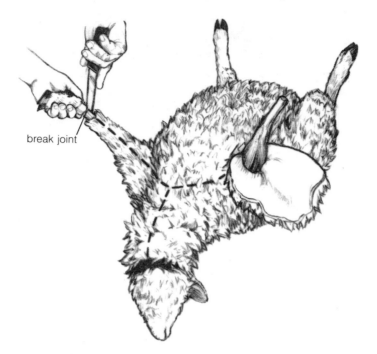

break joint

Figure 5.3: After the sheep has been bled, skin out the front legs. Remove the legs at the knee, or better yet at the "break" joint between the knee and ankle. Cut the skin from the inside out down the inside of the legs toward the front of the brisket and peel the skin back as far as the elbow. Make another cut perpendicular to the leg cut, from the brisket up the neck to the bleeding cut.

inside of the leg toward the front of the brisket and peel the skin back as far as the elbow. Now do the same on the other front leg and connect the two incisions just ahead of the brisket (**figure 5.3**).

Skinning and Removing the Organs

Cutting from the inside out, open the skin on the neck down to where you cut the throat. Using your clenched fist instead of a knife to separate the skin from the body, "punch" or "fist" the hide loose over the brisket as far back as the navel. Go to the rear legs and, starting at the dewclaws, skin each leg back to above the hock. Next, cutting from inside out, make an incision as far as a point just ahead of the anus. Skin the rear leg down to as far as the stifle, being careful not to cut the fell (the shiny thin membrane separating the muscle from the hide). Now fist the hide off the inside of the rear legs and down toward the navel (**figure 5.4**). At this point you may remove the rear feet at the break joint, the last joint above the hoof.

Go back to the front end and fist the hide from the brisket, around the navel, and back to where you worked in from the rear. Now hang the carcass, suspended by the hind legs, using a stout cord, light rope, or gambrel stick. Open the hide from inside out along the midline, loosening the navel as you do (**figure 5.5**). Fist the hide loose over the shoulders, and back and as far up as the tail (**figure 5.6**). Using your knife, skin around the tail and anus. You should now be able to drop the hide to the neck and peel it off completely. With your knife or, in older animals, the saw, split the breastbone its full length, being careful not to cut too deep and puncture stomach or intestines.

Figure 5.4: Skin the rear legs in the same manner as the front legs, and then fist the hide off the rear legs toward the navel. You may now remove the rear feet at the break joint.

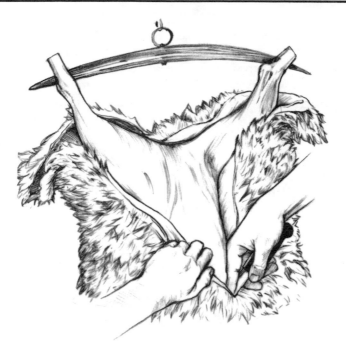

Figure 5.5: Hang the carcass by the rear legs from a rope or gambrel stick before the skinning and butchering processes begin. Open the hide from the inside out along the midline.

Cut around the bung, deep into the pelvis, and tie the rectum off so manure will not spill out (**figure 5.7**). To do this it will be necessary to pull the bung out of the pelvis, and it is easier to have a second person tie the string.

With a heavy knife or saw split the breastbone (**figure 5.8**). Being extra careful not to puncture the intestine, open the abdominal cavity just ahead of the pelvis. You will have to split the udder or cod (fatty tissue where testicles were removed) to reach this spot. As you open the cavity air will rush in and give you more room. Using your fingers as a guide, make this opening large enough to admit your hand. Again using your fingers as a guide, or holding the handle of the knife with your hand in and the blade out, carefully cut down as far as the breastbone (**figure 5.9**).

The paunch and intestines will now hang out. Be careful that the paunch does not break loose at the gullet and spill its contents. Reach up and find the bung and, using your

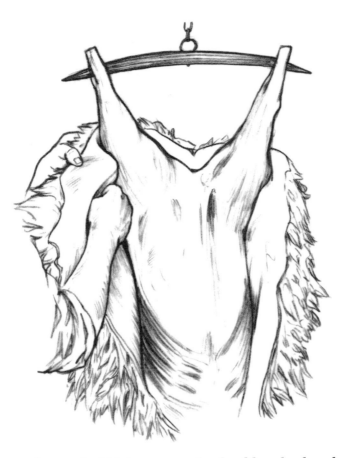

Figure 5.6: Fist the hide loose over the shoulders, back and as far up as the tail.

Figure 5.7: Cut deep around the bung, and tie it off with a piece of string.

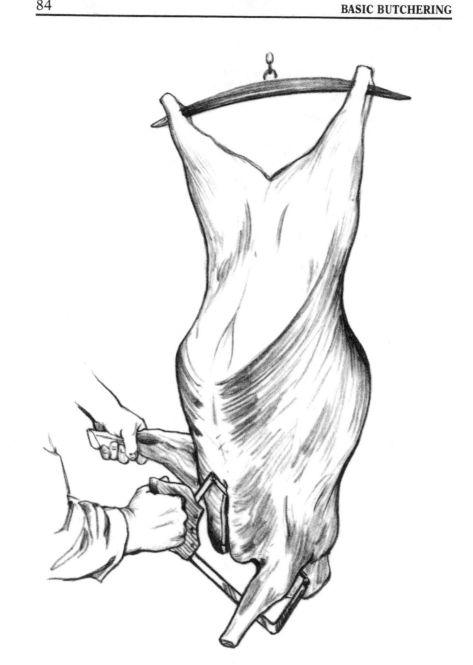

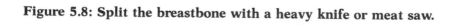

Figure 5.8: Split the breastbone with a heavy knife or meat saw.

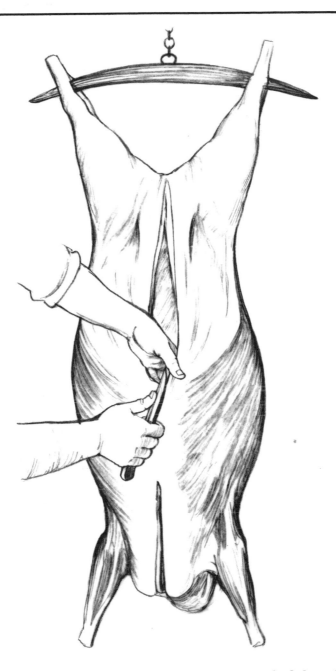

Figure 5.9: Open the abdominal cavity just ahead of the pelvis with a cut large enough to admit your hand. Take care not to puncture the intestine. Using your fingers on the inside as a guide, carefully cut down as far as the front end of the breastbone.

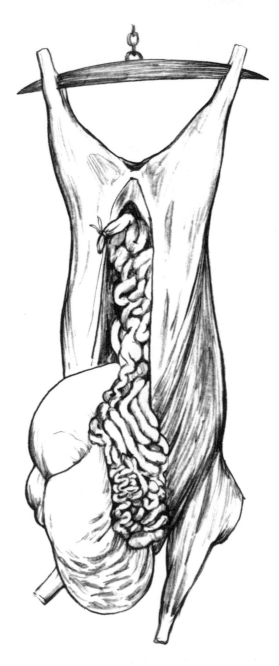

Figure 5.10: Reach up to find the bung, and pull it down with your left hand and hold it (if you are right-handed). With the knife in your right hand, loosen the intestines from their attachments.

fingers, loosen it and start to pull it down with your left hand (if you are right-handed) (**figure 5.10**). Use your knife to loosen the intestines from their attachments, but do not remove the kidneys. Remove the liver by cutting around behind it. Dissect the gall bladder out, rinse the liver, and hang it. When you reach the diaphragm, or skirt, cut around it and loosen the remainder of the "pluck" (lungs, heart, gullet, and windpipe) by cutting the large blood vessel on the top of the chest cavity (aorta). Follow the windpipe and gullet down to the place where you cut the throat, which should free the whole mass.

Salvage the heart by cutting it loose from its sack (pericardium). Split it open to remove the blood clot, rinse it, and hang it to cool. While you are salvaging sundries, grasp the stump of the windpipe left in the head and, cutting on either side, remove the tongue, a delicacy which may be boiled, pickled, salted, or smoked. Remove the kidneys and kidney fat, saving them for stew or to broil with chops for a mixed grill.

Butchering

Wash the inside of the carcass and the outside, too, if soiled, before moving the carcass to the chilling location. If the temperature is 28° to 40° lamb carcasses may be hung in a well-ventilated shed. Cover the carcasses with a sheet to prevent them from drying out. Do now allow carcasses to freeze. Although 24 to 48 hours at below 40° is said to be time enough before cutting and freezing a lamb carcass, I feel it is better to age lamb for about a week if the temperature is favorable. However, if you do not have a properly cool place and the weather turns warm, don't take a chance. Get it cut up.

5.A. Starting at the cod or udder, cut the flank loose as shown, going across the ribs and shoulder to just above the shoulder joint. Use the saw where there is bone.

5.B. Remove the foreshank from the breast and cut the breast into pieces (1), roll (2), and riblets (3). Remove the bone from the foreshank and use for Lamb Stew, or leave the bone in and use as Lamb Shank (4).

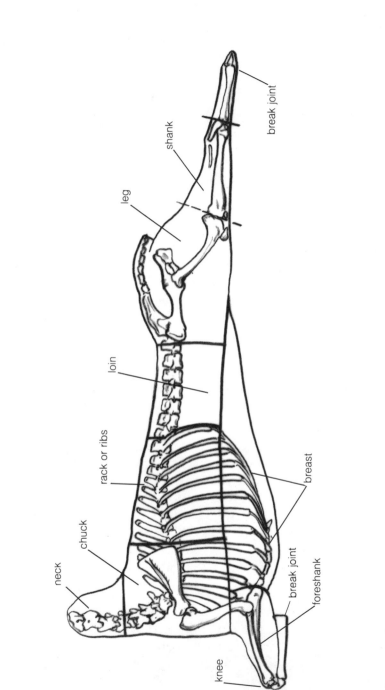

Carcass diagram.

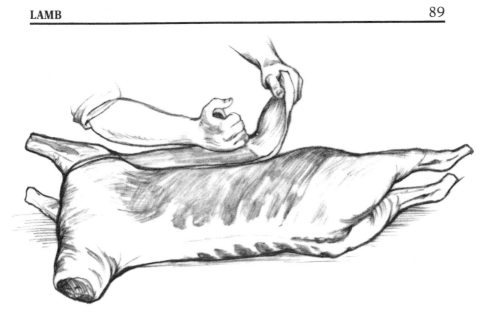

Figure 5.A: Cut the flank loose with a cut across the ribs and shoulder to just above the shoulder joint. Use a saw where you encounter bone.

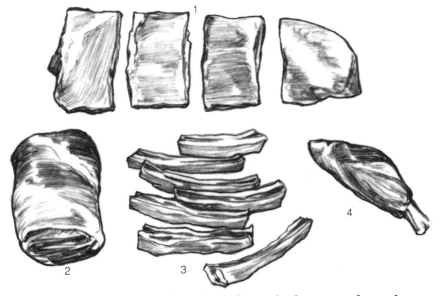

Figure 5.B: Remove the foreshank from the breast and cut the breast into pieces (1), boned and rolled (2), riblets (3), and foreshank (4).

As a boy I never thought much of sheep meat, perhaps because we only ate the old ewes and sold the good young ones. While in the army, however, I was invited to eat at a mess I was inspecting. The menu was braised Lamb Shanks, made from what was supposed to be the poorest cut of the lamb. The mess sergeant was a professional chef before the war, and he made that poor meat into a delicious feast. That experience taught me a lot and made me realize that no matter how good or poor the grade and cut of meat, the right preparation can make it a delicacy.

5.C. Cut just in front of the hipbone separating the legs (1) from the loin (2). Then cut between the last two ribs with a knife, and saw the backbone to separate the loin from the ribs or rack (3). Cutting at the level between the fifth and sixth ribs counting from the neck, separate chuck (4).

5.D. Separate the legs with knife and saw.

5.E. Prior to freezing prepare the legs for roasting by using your own judgment to make them the correct size for your family. For example, with the saw, cut three chops off the rump to make the leg smaller, or cut through the bone part way to make for easier carving. Cut the shankbone just below the stifle and either leave attached to the leg or remove and prepare as braised Lamb Shank. Legs may also be cut in two between stifle and aitchbone giving you roasts of 3 to 4 pounds.

5.F. After cutting the neck flush with the back, split the shoulder as shown. The neck may be boned for stew or ground lamb.

5.G. Use shoulder as roast (1), cut into chops (2) and (3), or bone and roll (4). To bone, follow bones with knife separating ribs first, then removing the arm (humerus), shoulder blade (scapula) and "back strap."

5.H. Split rack with saw as shown, cut into rib chops (see 5.I.), bone and roll, or prepare as Rack of Lamb.

5.I. Split loin the same as you did rack and cut into loin chops (1) and the rack into rib chops (2).

As you cut the lamb up, it is best to have someone else wrap, label, and put cuts into the freezer. If you work alone

Figure 5.C: Cut just in front of the hipbone to separate the legs (1) from the loin (2). Then cut between the last two ribs and backbone with a knife and saw to separate the loin from the ribs or rack (3). Count from the neck and cut between the fifth and sixth ribs to create the chuck (4).

Figure 5.D: Separate the legs with a knife and saw.

cut up only what you can wrap and freeze as you go. When preparing stew for freezing take time to remove all fat, sinew, and bone. Do not freeze anything that is not perfect. Lamb stew can be varied to make delicious meals — for example, a curry (see recipes).

Some people grind lamb and like lamb patties as well as beef hamburgers. The taste of these, too, depends on how well you trim out all undesirable parts. Even a tiny piece of spoiled, soiled, or too-dry meat ground with good will spoil the entire batch.

Figure 5.E: To prepare a leg for roasting, you may cut chops off the rump to make the leg smaller, and also cut part way through the leg bone to make for easier carving.

Figure 5.F: After you have cut the neck, split the shoulder with a saw as shown.

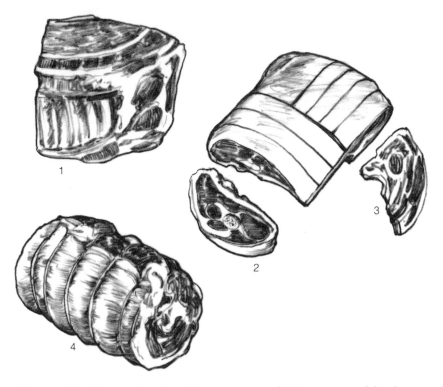

Figure 5.G: From the shoulder, you can obtain a roast (1), chops (2 and 3), or it can be boned and rolled (4).

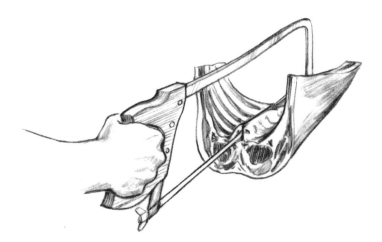

Figure 5.H: Split the rack with a saw.

Figure 5.I: Split the loin with a saw (as you did with the rack), and then cut into loin chops (1). You can cut the rack into rib chops (2).

6·VENISON

WHEN YOU FIGURE THE COST of growing it, the cost of harvesting it, and the amount spoiled and wasted through improper handling and butchering, venison is probably the most expensive meat produced in the country. If properly handled, however, it is some of the best meat you can eat.

In my area dairy farmers tell me they could keep 5 to 10 percent more cows if it weren't for the damage deer do to corn and alfalfa; fruit farmers also suffer severe damage to young trees from deer. Nevertheless, few want to see the deer herd reduced to pre-World War II levels. Deer were scarce then, but during the war the herd multiplied to a point where even with heavy hunting in the late 1940s and through the 1950s it grew to a level where it was eating itself out of food and shelter, and inbreeding was resulting in small, stunted animals.

During that same period two generations learned the joy of deer hunting, the pleasure of sitting quietly in the woods, and the good taste of venison. For a deer hunter the year is divided into three parts: the time before deer season, deer season, and the time after.

In the late 1950s our country had its first "doe day," a period late in the season when those who hadn't killed a buck could take a deer of either sex. Doe day began on a clear, cold morning, and the water pipes froze in the local auction barn, leaving a herd of cows without water. The man taking care of them hadn't gone hunting (he was seventy-six years old).

95

He figured the easiest way to water the cows was to turn
them out to a pond in back of the barn. One heifer, heavy
with calf, walked out on the ice and, when her rear feet slid
under her, sat down with a crash and went through the ice,
only her head remaining above water. The old man phoned
my office for veterinary help and the local fire company for
manpower to help pull the cow out. Since I had gotten a
good buck early in the season I was working and arrived at
the same time as the fire truck. In it were the only two other
men in town who were around to answer the fire siren: an
eighty-six-year-old and a man with a wooden leg.

At the suggestion of the older man, we laid a plank walk
on the ice to get to the cow, and chopped a channel with an
axe. The cow swam to shore. Had the siren signaled a fire,
the four of us could have done little more than watch it burn.

Over the years not only has the enthusiasm for deer hunting
changed, so has the philosophy. Some people now approve
of the once-scorned "meat hunter," and instead scorn the
person who hunts for sport.

Like many others, I consider deer hunting more than a
sport: it is almost a way of life. I enjoy watching deer year-
round and take pleasure in scouting for signs in the early fall,
when I pick up moss and partridge berries and evergreens
to make terrariums. When I hunt, preferably sitting quietly,
high on a ridge, watching the world of small creatures as
well as deer, the meat I may bring home is farthest from my
mind. Yet I don't criticize the unemployed man who shoots
a deer to feed his family or the overworked dairy farmer who
steps out the back door of the barn after morning milking
and drops a fat buck that has been feeding on his corn and
alfalfa all year. But I do scorn the person, whether a trophy
hunter or strictly a deer killer, who lets good meat go to
waste. Meat usually goes to waste from ignorance and misin-
formation rather than from neglect. Most people want to
salvage all the meat they can. This chapter is written for
those who want to do their own butchering and get the most
out of venison, both in pounds of good meat and satisfaction
in a job well done.

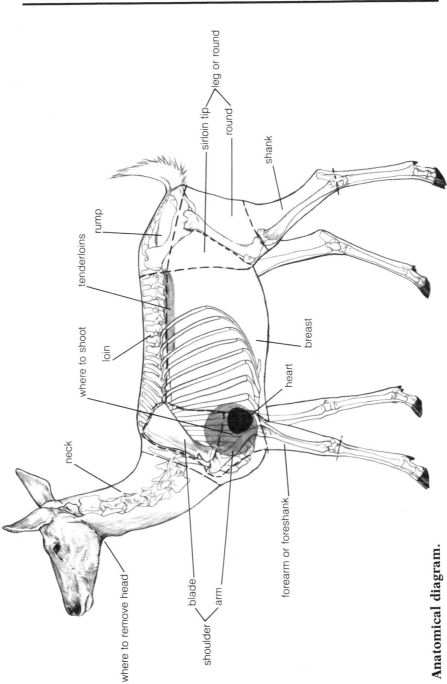

Anatomical diagram.

Most communities have professional butchers who will cut up deer. But some years when the weather is warm they can't handle all the deer killed in time to prevent spoilage. Besides, professional cutters don't have time to bone and properly trim venison. If you have the ability to kill a deer you can learn to bone and prepare it for freezing so as to have meat that will keep for a year without spoiling or taking on a bad taste. The trick is to remove all fat and bone.

Shooting and Dressing

First, to make good venison the deer must be shot correctly. That means a chest shot, preferably through the heart or close to it so good bleeding occurs. I know this is easier said than done, particularly for shotgun hunters, but even in that case, with modern guns and ammunition a good chest shot is not too much to ask. The neck shot, or "poacher's shot," is certain, either a kill or a miss; but by the time it is safe to "stick" the deer or cut its throat, the neck-shot animal won't be able to bleed out as well as one that has received a good chest shot just behind the shoulder, where good bleeding occurs inside the chest.

Field dressing, too, is vitally important. As soon as you are sure the deer is dead, and with it lying on its side, cut deeply around the bung (anus, or the anus and vulva in the doe) with your hunting knife. Pull the bung out so you can tie it off with a string to prevent its leaking manure when you remove it later with the rest of the insides through a midline opening **(figure 6.1)**.

Next roll the deer up on its back with the rear end downhill if you're not on a level place. Find the rear end of the breastbone (xiphoid cartilage) and make a hole through the skin just behind it. With the cutting edge of your knife up and pointing forward and down at a 45° angle, penetrate the belly wall, cutting forward and under the breastbone, being careful not to puncture the stomach **(figure 6.2)**. Withdraw the knife, insert two fingers into the hole to admit some air and, using your fingers as a guide, cut the belly wall midline as far back as the pelvis **(figure 6.3)**. At this time you may

Figure 6.1: With the dead animal on its side, cut deeply around the bung, pull it out, and tie it off with a string.

Figure 6.2: Roll the deer onto its back, and make a small cut just behind the rear end of the breastbone. Insert the knife at a 45° angle with the blade facing up, and cut forward under the breastbone. Be careful not to puncture the stomach.

Figure 6.3: Withdraw the knife and insert two fingers into the cut. Using your fingers as a guide, cut the belly wall along the midline as far back as the pelvis.

remove the testicles, but to comply with the game laws do not remove the sack.

Lay the deer on its side with belly downhill if feasible. The insides should roll part way out. Holding the belly wall open with one hand, use the other hand or your knife to break down the attachments of the insides along under the back **(figure 6.4)**. Cut the paunch loose where the gullet goes through the skirt (diaphragm), spilling as little of the contents as possible.

Now reach up into the pelvis from the front, and follow the large intestine (colon) breaking it loose as you go until you find the bung. Pull out the whole mass, which will usually include the bladder **(figure 6.5)**. Remove the liver, and set it aside on a clean rock or stump to cool until you put it in a plastic bag. Deer have no gall bladder, so don't be concerned that you have missed it.

The next step is very important. Go back around to the rear of the deer and with your knife ream the inside of the pelvis to remove all glandular tissue that failed to come out with the bung. That includes the prostate and seminal vesicles, which may give off a strong odor. Do not be concerned with the foul-smelling "brush" on the inside of the deer's hocks, or the scent glands near the dewclaws. You will spread

Figure 6.4: Now lay the deer on its side with the belly facing downhill. Hold the belly wall open with one hand, and with the other hand or a knife break down the attachments of the insides along under the back.

Figure 6.5: Reach up into the pelvis, and pull out the bung and the intestinal tract.

more odor by cutting them, contaminating your hands and knife.

If you only have a short drag to where you are going to hang your deer you may delay the next step. However, if it will be more than half an hour before you can hang the carcass, and if you have a good, stout knife, split the breastbone from the rear to the front, and from inside out **(figure 6.6)**. Hide on all animals should be cut from inside out to prevent getting cut hair on the meat, but in deer and sheep this is especially important. Cut the skirt parallel to the belly wall and remove the lungs and heart **(figure 6.7)**. If your knife won't cut the breastbone, cut the skirt and, reaching in with your knife, cut the lungs and heart loose. Save the heart and after it cools a bit put it in a plastic bag with the liver. Leave the kidneys in the deer. Roll the deer on its belly to a clean spot and, if there are two of you, one should take the horns (or ears), the other the tail, and "stand him up and shake him" to get all the blood out of the body cavity. If you are alone, lift one end at a time, starting with the head.

The classic drag rope goes on the horns, but one of my cattle clients showed me a better hitch years ago. He puts a rope calf halter on the deer and using a really short pull by putting a loop in the lead rope, he raises the head up, making it slide easier. Of course, if you can drive with the farm pickup or other vehicle to where the deer lies, you can wait to remove the heart and lungs until you get back to the place where you can hang the deer as you would a veal. Don't forget to put your hunting-license tag on the deer as required by law when you put the deer on the vehicle. If you have a long drive home to where you are going to hang the deer, spread the chest and belly open with sticks and carry it on the roof or tailgate of your vehicle, where it will get plenty of air.

Hanging and Skinning

Once back to where you are going to hang the deer, cut under the flexor tendon on the back of the hind leg between the dewclaws and hocks to raise the gambrel, just as you would on a hog (see **figure 3.5** on page 45). Don't hang the

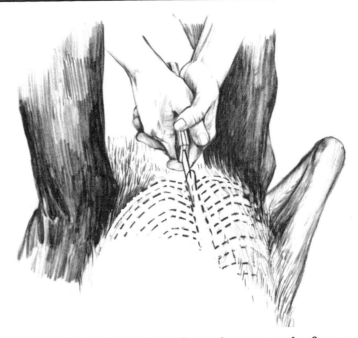

Figure 6.6: Split the breastbone from the rear to the front, and from the inside out.

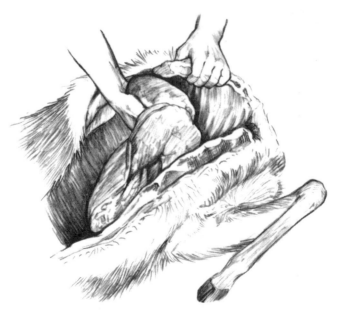

Figure 6.7: Cut the skirt parallel to the belly wall, and either pull out the heart and lungs, or cut them loose.

deer from the hocks as you would a beef, since that makes it more difficult to remove the skin. Put in a gambrel stick or pipe and raise your deer rear end up. Spread the body open with two sticks and remove the gullet and windpipe as far down as the deer's jaw, but if you're saving the head for mounting do not remove the gullet and windpipe until the skinning is completed. If plenty of clean water is available you may now wash the inside of the carcass. Never use water in the woods or before the deer is hung, since it won't drain out well and will cause spoilage.

In warm weather hang and skin the deer immediately and find a cooler to put the meat in for 24 to 48 hours before you cut and freeze it. In an emergency, split, skin, and quarter the deer and cut into pieces that will fit in an ordinary refrigerator. Once it's cool partially cover the skinless meat with plastic or a sheet to keep it from drying out and plan to get it cut up and frozen within 24 to 48 hours.

Skinning is easiest when the carcass is warm, but if the night-time temperature is from 26° to 40° and the day doesn't go above 50°, leave the hide on and hang the deer in a cool, airy shed or on the north side of a building out of the sun. Regardless of where you hang it, be sure it is out of reach of dogs and cats. Do not allow the carcass to freeze, particularly during the first 24 hours. If it does freeze solid, it may hang for up to three weeks frozen. Then bring it inside to thaw, and skin and cut it up immediately. Of course, if the carcass freezes and then warms up to over 50°, which it often will in November or December, skin and cut it up regardless of how long it has hung.

Carcasses that are badly shot up should be skinned immediately. Cut out damaged parts completely, even if that means destroying some good meat. The rest should be cooled below 40° and after 24 hours cut up and frozen.

To skin a deer, it must be hanging, either head up or head down depending on which is easier for you. The following description is for skinning with the head hanging down. There is really only one important thing in skinning technique, and that is to make all cuts from the inside out so as to avoid cutting hair and having it contaminate the meat.

First, cut around the cannon bone just below the hock (toward the hoof). Next, make a cut on the inside of the leg from this circular cut all the way to the place where you removed the anus. Now, still cutting from inside out, cut on the midline to where you made your initial midline cut to gut the deer.

Starting back up at the hocks, peel the hide toward the deer's groin. Be careful while doing this to avoid cutting through or handling the foul-smelling "brush" of light-colored hair on the inside of the hock. If you do handle it or get it on your knife, wash your hands and knife with warm soapy water to prevent carrying this odor to the meat.

If you are going to have the hide tanned with hair on, skin as soon as possible after killing and try to save the tail by splitting the skin of the tail all the way to the tip along the under surface of the tail. Although it is easier to cut the tail off and discard it, the white tail hair is used in fly tying to make certain types of flies, and hides with tails left on are worth a bit more to the tanner, even if hair is to be removed from the rest of the hide.

Once you've skinned out the hind legs and gone past the tail on a warm deer, you can usually pull the hide off as far down (forward) as the shoulders with little trouble by simply pulling and using your elbow as a lever **(figure 6.8)**. The longer the deer was hung and the colder it is, the more difficult the job, and the more you'll have to use your knife.

To skin the forward parts of the deer, cut from inside out on a line from the midline along the back of the deer's forelegs, over its elbow and on down to the forward cannon bone. The deer's "armpit" is a difficult place to skin and is often full of blood clots from the shooting. Take your time and do a clean job, trying to save as much meat as possible but being careful not to cut your fingers or the hide. Pull the hide down the legs and cut it loose at the cannon bone.

If you are saving the head of your deer for mounting, be very careful as you skin from the shoulders down the neck toward the head. Also, don't make a midline cut beyond the forward part of the deer's breastbone, and be very careful as you peel the hide from the neck. If you are not saving the

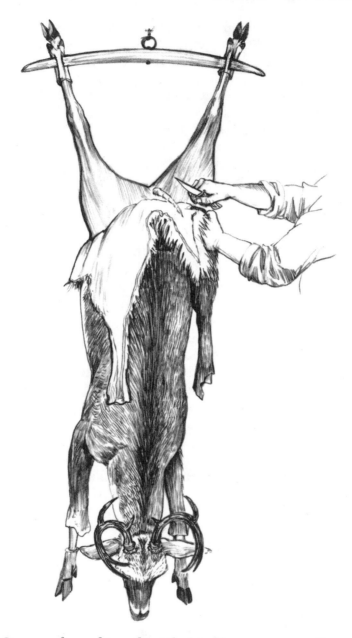

Figure 6.8: Once you have skinned out the hind legs, you can pull the hide off the shoulders by using your elbow as a lever. Cut and pull the hide from the inside out to prevent soiling the meat — this is especially important for deer.

head, continue the midline cut, peel, cut, and peel until you reach the head; then dislocate it at the first joint where the skull joins the first neck bone. Until you've done this a few times, you may find it easier to use a saw to remove the head. To avoid getting loose hair on the meat, don't cut the hide loose from the head until the head is removed.

To skin a deer hanging head up, start with a circular cut just below the ears and reverse the above procedure. When saving a head for taxidermy, you must skin with the head down so as to save as much hide as possible for the cape. Taxidermists want all the hide from the shoulders forward to make a realistic mount.

Regardless of when you skin, rub 2 to 3 pounds of salt on the flesh side of the hide, roll it, and hang it where the dog won't get it. The hide will sell for about $5 — not much of a price, but at least the hide doesn't go to waste. (Look for deer-hide buyers' ads in your local newspaper.) If you can find a place to have it tanned there is great personal satisfaction in wearing a vest or gloves from deer hide you "harvested" yourself.

The liver from a young deer is comparable to veal and can be cut in thin slices and fried or broiled (also see recipe for paté). The heart can be prepared as you would beef heart, and the kidneys should be handled like lamb kidneys. For a real gourmet treat remove the tenderloins as soon as your deer is hung. They are under the kidneys up along the backbone, and can be pulled out easily **(figure 6.9).** Cut into tiny steaks the size of a half dollar. Sear in butter and serve on buttered rolls with hot coffee — a delicious way to celebrate your successful hunt.

To freeze venison and have it keep well, bone it out and remove all fat, sinew, and nonmuscle tissue. The job is easily done with a sharp knife by anyone with enough do-it-yourself ability to kill a deer. It takes much longer than cutting with a saw as a professional butcher does, but the resulting meat is well worth the time and effort. Boneless, fat-free venison, well wrapped, will keep in a freezer literally for years, although game laws in most states require it to be used in less than a year.

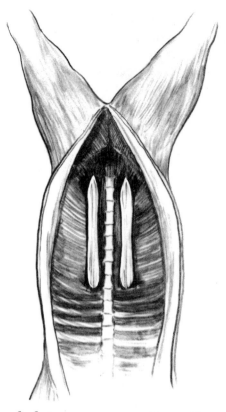

Figure 6.9: The tenderloins are a gourmet treat. You will find them under the kidneys up along the backbone. They can be pulled out easily.

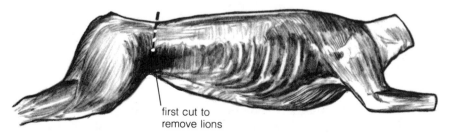

first cut to
remove lions

Figure 6.10: After you have completely skinned the carcass and removed the legs at hock and knee, spread the carcass out on its belly. The first cut you make will be to remove the loins. Cut just a little ahead of the hips down from the backbone on each side to the flank as shown.

If you don't wish to bone your venison see the chapter on veal and follow the butchering directions there — except you should make more cutlets (steaks) from your deer, and, as suggested in cutting veal, you should make lots of little scallopini steaks. Venison is fine processed as veal, but do not plan on storing it long with bone and fat left in. These contribute to the gamy taste many people don't like.

Butchering

When you have a skinned deer carcass with legs removed at hock and knee, place it spread-eagled on its belly on a wooden table high enough to let you work comfortably. Find the hipbone on each side, about 2½ inches out from the center of the backbone. With a sharp knife make a cut at right angles to the backbone just ahead of the hips. The cut should extend from the backbone to the flank of the carcass **(figure 6.10)**. Then cut as follows.

6.A. Cut as close to the backbone as possible from hip to shoulder, loosening the long loin muscles from its center attachment with your fingers. Now peel this out, cutting only on the ends and along the outside where needed. You'll now have a piece of meat about 20 inches long and 2 or 3 inches in diameter. Trim it of fat and sinew, cut in slices ¾-inch thick, wrap, and label boneless loin chops. Do the same on the other side.

6.B. Remove a shoulder starting at the brisket, cutting up under the "armpit" and staying as close to the ribs as possible until you reach the top of the withers, freeing the whole shoulder.

6.C. Cut across the shoulder at the level of the shoulder joint and sever the joint. You could save this piece as a roast with the bone in, but for proper storage dissect the shoulder blade out, starting at the joint and working toward the top. You can trim, roll, and tie it as a Boneless Blade Roast, or separate the muscle bundles on either side of the "spine" of the shoulder blade. Slice them as boneless shoulder chops or thin little scallops.

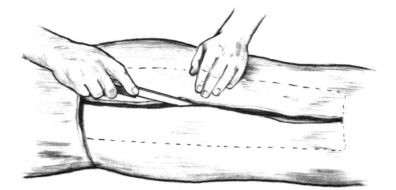

Figure 6.A: Cut as close to the backbone as possible from the hip to the shoulder. Loosen the loin muscles with your fingers. Cut as shown to obtain two pieces of meat that are each approximately 20 inches long and 2 or 3 inches wide.

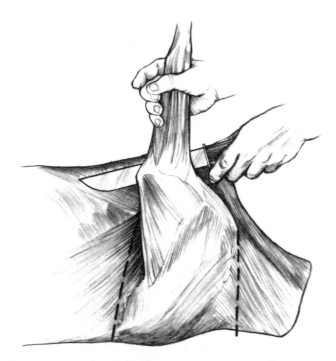

Figure 6.B: Remove the shoulders by cutting from the brisket to the armpit. Stay as close to the ribs as possible until you reach the top of the withers. At this point the whole shoulder should come free.

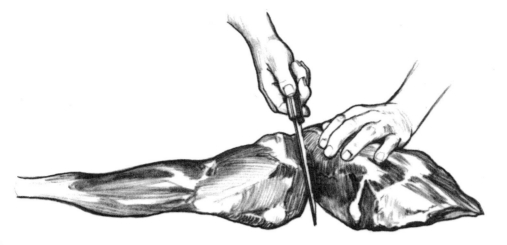

Figure 6.C: Cut across the shoulder at the shoulder joint and sever it.

6.D. Cut at the elbow joint to separate the arm and shank. You may saw here, using a carpenter's saw if you don't have a meat saw. If you have neither, separate the joint with your knife. Trim the armbone (humerus) out and roll the meat for a roast or cut for Swiss steak, stew, or chopped meat. The shank may easily be boned and the meat used for stew or grinding. Don't overlook the possibility of slicing some for scallops.

6.E. Taking your time and using your ingenuity, bone out the neck and roll it for a roast. The neck makes excellent sauerbraten (see recipe on page 160). You can also saw it off and cut it up for stew or use it for grinding.

6.F. If you have a saw, cut the backbone off just ahead of the hipbone where you made the first cut to remove the loin, and split the legs at the midline as shown. Or you may turn the carcass on its back and, spreading the legs, cut down until you find the large ball-and-socket joint and remove the legs. Even if you split the pelvis with a saw, remove the legs at this joint.

The meat left on the pelvis is the rump. With a sharp knife, by following the pelvic bone, you may peel away a small piece of tender dark meat usually covered with a heavy layer

Figure 6.D: Cut the elbow joint to separate the arm and the shank.

Figure 6.E: Cut off the neck with a saw, and then bone it out to obtain roast meat.

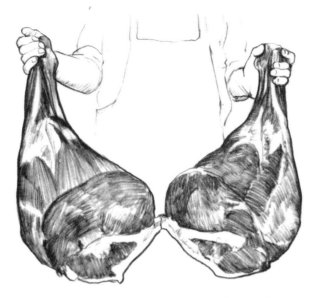

Figure 6.F: Saw off the backbone at the place where you made the first loin cut. Split the legs at the midline as shown.

of fat. Remove the fat and roll the meat as a roast or cut it into thin steaks or scallops. Trimmings go for stew or grinding.

The best meat you have left now is the ham, consisting of sirloin tip, round, and shank. Here you may wish to refer to the chapter on veal and make large round steaks (called cutlets in veal) as shown there. The legbone and extra fat should be removed before freezing.

6.G. Remove the sirloin tip as shown by starting at the stifle and cutting along the long bone (femur).

6.H. Trim the sirloin tip into one football-shaped piece that makes good steak, Swiss steak, or scallops, and use the trimmings for stew or grinding meat.

6.I. Separate the femur from the meaty part of the round, leaving the meat below the stifle on the shankbone (tibia). Trim the meat off the shankbone and use for stew or grinding.

6.J. Follow the natural separations in the round, and you will come up with two large pieces of muscle. These may be used as roasts, but are usually cut as round steaks. If you don't separate the muscles you may slice larger steaks.

Figure 6.G: Remove the sirloin tip by cutting from the stifle along the long bone or femur.

sirloin tip

trimmings for stew or grinding

Figure 6.H: Trim the sirloin tip into one football-shaped piece of meat and use the trimmings for stew or grinding meat.

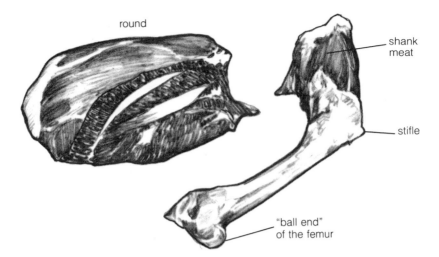

round

shank
meat

stifle

"ball end"
of the femur

Figure 6.I: Separate the femur from the meaty part of the round, but leave the meat below the stifle on the shankbone (tibia). Trim the meat off the shankbone and use for stew or grinding.

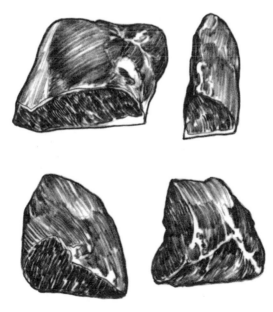

Figure 6.J: Follow the natural separations of the round, and you will obtain two large pieces of muscle. This muscle may be further trimmed into four roasts as shown, but are usually cut as round steaks.

Now you have only the backbone and ribs left. On a particularly large deer you may find enough meat on the ribs to bone and use. Rib meat is not usually very good, but it will be enjoyed by your dog. Don't be afraid to feed venison bones and scraps to a dog. A dog is no more apt to chase deer because it has eaten venison than to chase cows when you feed it beef.

Wrap venison extra well, as you do veal, since it has no protective coating of fat and will get freezer burn more readily than beef. Venison may be ground as hamburger, but usually pork or beef fat is substituted for the venison fat, which becomes rancid quickly.

Venison makes excellent jerky and also may be salted and smoked Norwegian style (see pages 146–48).

7·POULTRY

CHICKEN IS THE MOST POPULAR of the various species of poultry eaten, so we shall discuss the use, slaughter, and butchering of chickens first, then explain the differences in handling between chickens, ducks, geese, turkeys, and game birds.

For food purposes, chickens are classified as follows:

TYPE	AGE	WEIGHT
Broiler	8–14 weeks	1–2½ pounds
Fryer	14–20 weeks	2½–3½ pounds
Roaster	5–9 months	Over 3½ pounds
Capon (de-sexed male)	6–10 months	Over 4 pounds
Fowl	Over 9 months	Over 3 pounds
Roosters	Over 10 months	3–6 pounds

My boyhood memories of picking chickens at home on the farm are not pleasant. It was a job I didn't enjoy, like weeding carrots and picking peas. Still, no chicken prepared by a gourmet chef or at a country barbecue ever tastes as good to me as the broilers my mother cooked on the wood stove along with the peas I hated to pick and fried new potatoes I enjoyed digging. You can probably buy chicken at a local market cheaper than you can grow it, but it will never taste as good.

Select healthy, fast-growing birds to slaughter and, if you haven't done it before, start out with only two or three on the first try. Separate the ones to be killed so they may be starved for 24 to 36 hours to empty out the digestive tract.

117

That makes for cleaner, easier "drawing" (eviscerating). Allow the birds free access to water during this starvation period.

To butcher chickens you will need a sharp knife with a narrow 3- to 3½-inch blade such as a jackknife. A "hoe-blade" castrating knife may also be used if it is sharp. Baling twine, or better yet light nylon cord, is needed to hang the birds and, of course, you need a place to hang them at a convenient height. For a less messy killing area "blood cups" can be made out of soup cans, attached to the beak of the bird with a wire hook after sticking, and weighted with a piece of iron on a string to keep it and the bird from flopping. You will need pails and a source of heated water. A barrel or garbage can to pick feathers into is handy. A propane torch should be available to singe hair off plucked birds, and a blunt knife to make picking pinfeathers easier under a cold water faucet is useful.

To butcher you will need at least one sharp, narrow-bladed knife such as a boning knife. A source of cold water is needed both to wash birds and to prechill them. Ice should be available to chill water.

Killing and Plucking Chickens

To kill a chicken, hang it by its legs and grasp the head with the comb in your left hand. Insert the knife back into one side of the throat and cut to the opposite side as you withdraw it (**figure 7.1**). You will know you cut major blood vessels when blood gushes out. Now find the groove in the roof of the mouth and force the knife into the rear part of the bird's skull, piercing the brain (**figure 7.2**). A squawk will indicate you have done the job. Although the bird will be dead the heart will keep beating and pump blood out. Hang a blood cup to catch the blood and keep it from splashing.

The foregoing is standard advice on killing chickens, but I prefer to reverse the sequence, sticking the brain first, then cutting the jugular vein, simply because it seems more humane. Of course, if you chop the chicken's head off with an axe or hatchet you will do the job quickly, but it is messy.

When you are butchering only a few birds, dry picking is the best method of removing feathers, and it yields a better-

Figure 7.1: To kill the chicken, hang it by its legs, insert the knife back into one side of the throat and cut to the opposite side as you withdraw the knife.

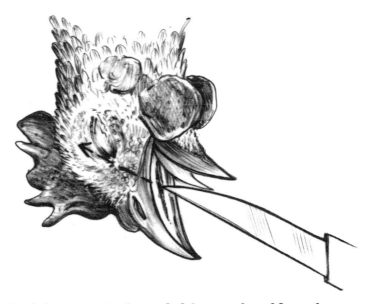

Figure 7.2: Find the groove in the roof of the mouth and force the knife into the rear part of the skull, which will pierce the brain.

looking carcass. It must be started as soon as the bird is dead and will not work unless the bird is properly stuck. Hold the bird by the neck and start picking in the direction in which the feathers lie over the breast, taking a few feathers at a time so as not to tear skin. Next do the neck and the body, finishing with the wings and tail.

To pick wet by scalding, dip the dead birds one at a time into water at 128° to 130° for 30 seconds. The water need not be as hot for young birds as for older ones, but the time is the same. Move the bird up and down in the water to soak through the feathers onto the skin.

Pinfeathers come out most easily under a stream of cool water. You may rub them out and finish removing the few more difficult ones with a dull knife against your thumb, or a tweezers. If the bird has hair, singe with a propane torch, directing the flame parallel to the skin. You can do it over a gas range if you don't have a torch.

Cutting Up

Now remove the head with a sharp knife or shears through the joint between the head and neck. With the bird hanging, remove the oil gland on the back just ahead of the tail. Cutting from in front of the nipple of the oil sack toward the rear, scoop it out as you cut (**figure 7.3**). Be sure you get it all. Cut feet off at the hock joint, and you are ready to draw.

With the bird on its back, cut through the skin at the point of the shoulder, and continue the cut up the back of the neck to where you took the head off. Remove the crop, windpipe, and gullet. Now, by first cutting the muscle and then twisting, remove the neck close to the body.

Carefully make a cut just under the rear of the keelbone (breastbone). Continue the cut down and around between the anus and the tail and back up, trying not to cut the intestines (**figure 7.4**). Hold the anus in one hand and with the other go in under the keelbone, breaking down attachments, and scoop the insides out. You may have to go in a second time to get the gizzard.

Remove the liver and trim out the gall bladder. Trim the heart and save it. Cut the gizzard under flowing water

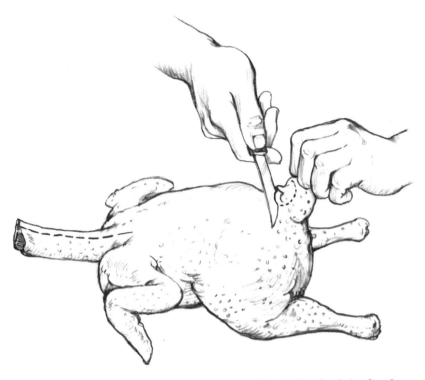

Figure 7.3: After you have scalded the bird and plucked the feathers, remove the oil gland on the back just above the tail. Cut in front of the nipple of the oil sack toward the rear, and scoop it out.

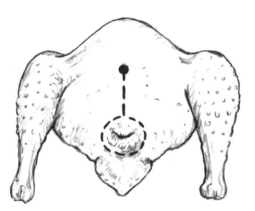

Figure 7.4: To prepare the carcass for eviscerating, make a cut just under the rear end of the breastbone down to the tail and anus. Cut around the anus as shown. Hold the anus in one hand and scoop the insides out with the other hand.

through its thin side and peel out its lining. Rinse the liver and heart and put in cold water to chill with the gizzard and neck.

Remove the gonads (ovaries or testes) from up under the back and scrape the lungs out with a tool or your fingers. Using a hose, or under a faucet, rinse the carcass inside and out. Put the carcass, or carcasses, in a container and let clean tap water run in, fill, and overflow for a time long enough to remove body heat.

Then put carcasses in a tub of ice water and leave them there until they have reached a temperature of 40° or below, which will take nearly 3 hours. Pull the carcasses out and hang to drain for 10 to 20 minutes.

Now dry and wrap the neck and giblets in a waterproof sack and put inside the bird. (These make good soup, or can be cooked and chopped to add to gravy.) Birds may be refrigerated at 29° to 34° for up to 5 days before being consumed, or up to 3 days before being wrapped and frozen. Do not attempt to freeze poultry until it is down to at least 40° or below. Birds should be well wrapped to prevent freezer burn. In fact, tightly closed plastic bags are better for poultry than freezer wrap.

Turkeys

Turkeys are handled in much the same way as chickens, but require more and hotter water to pick. Dry pick if you can, and have plenty of cold water and ice to cool the much larger carcass. Turkeys' legs have hard, bonelike tendons. Remove legs by cutting around the hock joint and breaking the joint open over the edge of a table, then pulling (**figure 7.5**). Do not attempt to freeze turkey or large chickens until they have cooled at least 20 to 24 hours.

Ducks and Geese

The main problem in preparing ducks and geese is caused by the water-resistance of the feathers and down. Again dry picking is by far the most desirable method of getting the

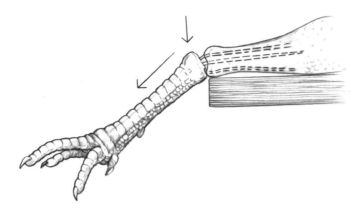

Figure 7.5: Remove a turkey leg by cutting around the hock joint, and break the joint open over the edge of a table. Then pull the leg loose from the hard tendons.

feathers off. If you must scald, have water at 135° and add a little detergent. Even so, it takes longer to loosen the feathers. Some people soak burlap bags in hot water and wrap the birds in them for 2 to 3 minutes to loosen feathers.

Regardless of how you pick ducks and geese, dry or wet, you will be able to have a nicer-looking carcass if you rough-pick first then cool the carcass and finish with paraffin. Heat enough water in a pail to dip the entire duck or goose, and make the water hot enough to melt two ¼-pound slices of paraffin. Dip the somewhat cooled bird for 5 to 10 seconds, pull it out, and hang in a cool place for 2 to 3 minutes. The paraffin will have cooled in a layer on the carcass. Pull it off. With it should come the remaining feathers, and any dirt or soil left on the carcass. On occasion you may have to repeat the process or pull a few pinfeathers by hand.

Game Birds

Game birds and wild waterfowl are handled similarly to the domestic variety, with a few exceptions.

Pheasant, grouse, quail, wild doves, and woodcock are most easily plucked using hot water. Temperatures of 126° to 128° should be hot enough, and the birds are immersed only long enough to loosen their feathers. During warmer weather draw the birds in the field as soon as they are shot. Whenever possible, hang them in a cool shady place or get them into a cooler as quickly as possible.

Older literature says game birds and wild waterfowl should be aged with the insides left in. Modern taste does not appreciate the gamy flavor that may result, however. To my way of thinking the gaminess is simply the taste of the intestinal contents of the bird, which goes all through the meat. After all, most birds shot have punctured intestines. The quicker you clean them and get the carcass chilled to stop bacterial growth, the better. If properly cared for after shooting, ducks and geese feeding on corn all fall don't taste any different from corn-fed domestic birds.

It has been my experience that wild turkey, ducks, and geese may be picked dry if the work is started within 2 hours after shooting. The young fall turkey requires a lot of pinfeather picking under the cold-water tap. Waterfowl, if rough-picked dry, clean up well using the paraffin method described earlier in this chapter.

8·RABBITS AND SMALL GAME

RABBITS ARE THE EASIEST home-raised, meat-producing animal to slaughter and prepare. The meat is delicious, and there is no end to the variety of ways it may be served. For cholesterol and/or weight watchers, rabbit is a tasty low-fat meat.

Young rabbits 2 to 6 months old seem to be the most desirable, but those of any age may be butchered. If one is too old to make frying meat it should still make good hasenpfeffer.

Killing and Skinning Rabbits

To kill a rabbit hold it by the hind legs and hit it behind the ears on the back of the neck with a club (**figure 8.1**). The head should be removed immediately with a sharp knife so the rabbit will bleed out.

As soon as the rabbit is dead, hang it by one gambrel (hock) joint on a number 6 screw hook on a beam at a convenient height for working on the carcass. With a knife remove the other rear leg at the hock joint, and the front feet and tail. Skin the free leg back to the region of the anus and groin using your fingers, the knife being used only to start. Skin the other leg (the one hooked at the gambrel) down to the groin and then, without using the knife, peel the whole hide down and over the body as you would turn down a sock (**figure 8.2**).

Being careful not to puncture intestines, make a slit the full length of the body from groin to front of breastbone

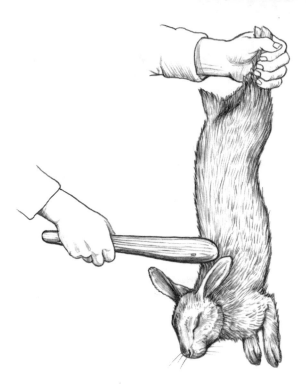

Figure 8.1: To kill a rabbit, hold it by the hind legs and hit it behind the ears on the back of the neck with a club.

(figure 8.3). Again using your fingers, by blunt dissection remove the entrails, leaving the liver, kidneys, and heart in. The carcass should be rinsed in cool water and cut into seven pieces as shown in the figure: 1-leg, 2-loin, 3-shoulder, 4-back **(figure 8.4)**.

Chill the meat in ice water for 2 hours, then wrap and freeze. Rabbit meat may also be stored at 29° to 34° for up to 5 days and used fresh. Rabbit may be prepared in many ways. See recipes on pages 178-80.

Wild rabbit may be skinned and prepared in the same way as the domestic animal. However, if you're butchering a wild rabbit that was shot during warm weather, remove the insides immediately and leave the skin on until you reach a place where it may be skinned and chilled. Of course cut away all shot, damaged parts.

Figure 8.2: Hang the animal by one hock on a screw hook, and remove the other rear leg at the hock joint, the front feet, and the tail. Start to skin the carcass with a knife, but then peel the entire hide down over the body.

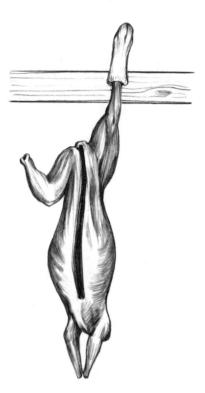

Figure 8.3: To eviscerate the carcass, make a cut down the full length of the underside from the groin to the front of the breastbone. Use your fingers to dissect and remove the entrails.

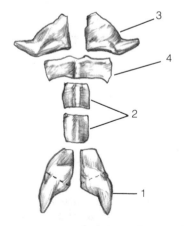

Figure 8.4: Rinse the carcass and cut as shown: leg (1), loin (2), shoulder (3), back (4).

Squirrels, Raccoons, and Woodchucks

Squirrels are more difficult to skin than rabbits, and you will find you must use your knife more. As in all game, the quicker you remove the insides the better the meat will keep and taste.

Raccoon is delicious if properly prepared, the most important step being removal of all visible fat from the carcass. Animals less than a year old are probably the best for eating, and should be skinned and gutted as soon as possible after killing. The carcass should be washed and hung to cool immediately. The fat may be removed easier once the carcass is cool than when it is fresh, but regardless of when you butcher a coon, take your time, and get all the fat out. See page 178 for recipes.

Woodchucks are one of the cleanest animals, eating only grass (or alfalfa or clover), and if properly prepared are tasty. They must be skinned and cleaned immediately on being killed. The secret to using them for human food is to remove a tiny dark gland under each front leg that will give a bad flavor. As with raccoon, remove all the external fat that you can. See recipe on page 178 for use of young woodchuck.

9·LESS POPULAR MEATS

WITH THE POPULARITY of dairy goats increasing in recent years, a problem has arisen as to disposing of young males. Those born up to 3 months prior to Easter find a ready market the week before Easter as "baby goat," but the rest of the year it is often considered economically wise to put the new-born billy painlessly to sleep rather than to raise him and sell him six months later for two dollars.

Butchering Goats

Today many people have found that young goats 4 to 12 months of age, if castrated and fed properly, make excellent meat. Animals can be prepared as whole-carcass "baby goat," and they can be cut up the same as lamb or veal. The best way to cut up a goat, however, would be to bone and remove all fat and sinew as described in chapter 6 for venison. In this way the meat can be frozen and kept for a longer period of time without acquiring an off flavor.

Anyone who can butcher a veal, lamb, or deer can certainly butcher a goat. They are easier to skin than a lamb since goat hair doesn't give off a bad flavor, as wool does if one's hands become contaminated by it. Still, no matter what you butcher, be clean when you skin, and wash your hands of all hair, dirt, and contamination before you ever touch meat.

Milk-fattened goats should be the best, but if you keep a male goat over 3 months old for meat, castrate him and then,

131

once he's weaned, feed him good hay and grain for fast growth.

Of all the animals that are used for meat, the whole, or uncastrated, billy goat is the least edible. Even an uncastrated boar hog can be put into strongly seasoned, spiced, and garlic-flavored sausage, but the goat? Never!

Several years ago two local tavern owners in my area, although good friends, were fierce competitors for the game they hunted as well as for customers for their respective hotels. Andy knew where Bill's favorite deer stand was, so when he saw an old tan billy goat with long straight horns for sale at the local livestock auction, he got a mischievous idea and purchased the goat for the great sum of $2. The next morning as the fog lifted on Kiekout Mountain, Andy's rival, Bill, spotted a deer with great long spikes eating brush. The "buck" was easily shot, but when Bill climbed down from his tree stand and ran to his trophy he found a collar around its neck and the collar fastened to the tether tied to a buckleberry bush. There was no question in Bill's mind as to where this foul-smelling deer came from.

The following morning farmers delivering milk to the local milk plant saw an old tan billy goat hanging on the porch of the village inn with a sign beside it reading, *Look what I shot last night with my little jack light,* and it was signed, *Andy.*

By 8:00 A.M. the goat was gone, and never a word was said until a few weeks later when an out-of-state hunter stopped at the inn and said to the proprietor, "I tried to buy some venison from a hotel in the next town and they didn't have any, but they said a man here named Andy sells a lot of it." He explained that he needed it for a game dinner at his firehouse.

Andy tried to act insulted at the suggestion that he would sell venison, but he allowed as how the hunter appeared to be a good sort, and that since it was for a fire company he would give him some of his personal supply from the freezer. The hunter showed his gratitude by buying several rounds for everyone at the bar and left with his prize. As he left Andy cautioned him that since this meat was from a huge old buck it must be parboiled for several hours prior to other cooking.

Rumor has it that not long after these events a certain firehouse was burned down by the firemen. The place had developed such a bad odor on the afternoon of the annual game dinner that everyone who entered became ill, and even the most thorough deodorizing could not remove the bad odor.

So remember, never butcher a billy goat!

Horses

One other domestic animal you might consider butchering and eating is the horse. Other countries use horse meat interchangeably with beef, but although many horses go for slaughter in this country all but a small percentage go for export. I have eaten steak from a yearling horse that was put down because of congenital blindness. That was while I was at veterinary college, and the meal was more in the spirit of "I'll try anything once" than it was motivated by a desire for a meal. Having grown up with horses and, like most Americans, believing that there is a mystical bond between horse and man, I think I'd become a vegetarian if there was nothing else to eat but horse meat, not because of the flavor or quality of the meat, but just from sentiment.

Still, if a family had the choice of horse meat or no meat for the winter, I cannot argue with their choice to eat horse meat.

To butcher a horse one would have to shoot it in the head at the X where the lines from ears to opposite eyes intersect, and then follow about the same directions as for slaughtering beef. Cuts would be the same as beef, too, although as with the goat and deer, I would suggest getting rid of all the fat possible before freezing. For grinding horse meat, beef fat should be added instead of horse fat.

Horse meat tastes somewhat sweeter than beef and thus may need more seasoning. In Norway it sells in the market, salted and smoked in a sausage like salami. Again, the need for strong seasoning seems evident.

Buffaloes

The last meat animal we'll mention is sort of halfway between domestic and wild, the bison or buffalo. Beefalo, a cross between cattle and buffalo, are butchered exactly like beef. The full-blooded buffalo kept on farms for beef should be handled the same way. Its wild ancestry makes the buffalo difficult to handle for slaughter, and makes its meat lean and gamy.

Recalling what we said in chapter 1 about fright affecting meat, shooting a buffalo in the head with a high-powered rifle before it has a chance to become upset would seem prudent. Again, the place to shoot is not between the eyes, but where the lines of an X drawn from horn to eye intersect. Cutting the throat for bleeding should be done as usual, and to get good bleeding, it should be done as quickly as possible.

Buffalo meat should make excellent jerky, and it should take well to salting and smoking. If ground, beef fat or fat beef should be added to bring the quality of buffburger up to that of beef.

10·MEAT INSPECTION

WHEN YOU BUY MEAT at the store it has been inspected and approved at slaughter by a federal or state veterinarian and followed through processing by a qualified meat inspector, thus ensuring that the animal it came from does not have a disease such as tuberculosis that a human could contract, and ensuring too that the meat is not contaminated with disease-producing organisms. You are prohibited by federal law from selling meat slaughtered at home, with the exception of domestic rabbit. For your own protection, you should familiarize yourself with the Federal Meat Inspection Act, which appears at the front of this book.

How do you know the meat you butcher is safe for your own use? First, be sure the animal you butcher is healthy. Second, if you see anything that appears abnormal when you open an animal, call for veterinary advice. Third, be as clean at slaughter and butchering as you would be in your own kitchen. Fourth, be careful in preparation to discard every bit of unclean, bloodshot, too dry, too "slippery," or foul-smelling meat, and in preservation (cooling, freezing, cooking, salting, pickling, and smoking) don't take short cuts that permit disease organisms to get a foothold.

If you are going to butcher your own animals, chances are you are going to raise them yourself. If you are raising animals you should make the acquaintance of your local veterinarian *before* you need his or her services, so that when you need help — which might be to TB test your cattle or goats or to decide if an animal you are butchering is fit to eat — he or she knows who and where you are.

To avoid problems, butcher only animals that appear to be healthy. Of course many of the animals butchered at home are butchered in an emergency — for example, the fat two-year-old ready to calve that breaks a leg. It's a choice of twenty-five dollars and torture for the heifer to go to the dog-food plant some hours away, or an emergency butcher job resulting in five hundred dollars worth of good meat for you and your hired man. In such a case, call your local veterinarian and ask for advice.

If after you hang a carcass you find a lot of questionable meat or find an abscess or tumor, or anything else that doesn't look right, call your veterinarian who may be able to advise you on the telephone. If not, he or she will come out and decide whether the meat should be used.

In general, never butcher an animal running a fever or with diarrhea or any other sign of illness. Even a healthy-looking animal may have an abscessed liver, or an old "hardware abscess" between the heart and second stomach, caused by metal objects the cow has eaten. If that is the only lesion found, your veterinarian will probably say "cut it out, clean it up well, and use the meat but not the liver if that is where the abscess was." If there are multiple abscesses your veterinarian may tell you to bury the whole carcass. Calves, pigs, sheep, and goats may have spots on the liver from worm (parasite) lesions. If the rest of the carcass is all right you need only throw away the liver.

Older cattle may have lymphoma lesions and still appear healthy. These lesions look like fat tumors but are not. Like other lesions, one lymphoma is probably nothing to worry about and may be cut out and ignored. Two or more, however, indicate a general case, and you might do best to bury the animal.

Deer covered with wartlike tumors are said to be all right to eat, but who would want to? Wounded deer should be destroyed to terminate suffering but, for example, if a leg is infected and foul-smelling you had better not use the meat.

Rabbits with spots on their liver may be carriers of

tularemia. Be careful of even handling them if you have cuts on your hands or if the animals are covered with fleas. It's not dangerous to eat the meat of such rabbits, however, if it is thoroughly cooked.

In every case, the best advice I can give you is to call your veterinarian if you discover questionable meat. He or she is the most valuable friend a country person can have.

11·PROCESSING AND PRESERVING

MOST MEAT TODAY is preserved by freezing. The foregoing chapters have already given some detail on preserving specific meats. For example, pork does not improve by aging, so if it is to be frozen or preserved in any way, start the process withint 3 to 5 days after slaughter. This chapter will cover general procedures on cooling and freezing, and will give specific directions on basic ways of salting, pickling, and smoking, as well as sausage making. The directions given are only examples of techniques. Every home butcher has his own favorite recipes for preserved meat that his family likes. Collect as many different recipes for preserving and sausage making as you can, and vary the basic ones here to suit your taste.

Spoilage

Earlier in the book we mentioned sanitation and cleanliness, but we would like to reemphasize it here. Spoilage in meat occurs for two reasons: by bacterial breakdown, and by enzyme action. Both bacteria and enzymes must have warm temperatures to act (they work fastest at body temperature), but are also able to act at below-freezing temperatures over a period of time. In some foods, including meat, "good" bacteria and normal enzyme action are used to ferment and flavor food. Fresh meat will not spoil as quickly if it is not contaminated by bacteria from filth during butchering. Thus our emphasis on cleanliness and washing hands, as well as on washing the part of a carcass that becomes soiled by ingesta (contents of the digestive tract) during butchering.

Especially dangerous types of bacterial growth, particularly in preserved meat, but even in ground fresh meat and poultry, can cause food poisoning. Bacteria of the Salmonella and Staphylococcus genera are the culprits. Salmonella bacteria are common in the digestive tract of animals and poultry, and if introduced to food by the unwashed hands of a food handler, will grow at temperatures over 40°. If eaten by humans those organisms can cause a sometimes fatal diarrhea. Staphylococcus, or "staph," is most often introduced to food by food handlers with infected wounds on their hands. Staph will grow in food, die, and form a toxin which is poisonous to humans. Very serious and sometimes fatal diarrhea and vomiting will start within minutes or hours of eating food containing staph toxin.

Chopped meats and sausage products are particularly dangerous if they become contaminated by either salmonella or staph since they are often stored or worked on at temperatures above 40°. To be sure you never have a food-poisoning problem, with meat you butcher and preserve, follow two simple rules, making no exceptions: 1) Always wash your hands with soap and water and dry them before handling food and after every time you use the toilet; 2) Never allow anyone to work on food who has an open sore or wound, or who has diarrhea.

A third deadly bacterial disease, botulism, is caused by a toxin formed after the bacteria, *Clostridium botulinum* grows in food stored without air, such as in plastic bags, or cans or jars. Botulism is prevented by proper heat applied to food at canning and by the use of saltpeter (potassium or sodium nitrate) in salting and brine mixture. Saltpeter also gives preserved meats the characteristic red "ham" color. Don't let the possible link between cancer and nitrates (which come from the breakdown of nitrates at cooking under certain conditions) scare you into leaving the saltpeter out of your salt preserving mixture. Botulism kills in hours and is a definite possibility when preserved foods are not treated with sodium nitrate. By comparison, nitrate-caused cancer is only a remote possibility. Don't take chances on botulism.

As soon as an animal dies bacteria invade its system, and the only way to stop their growth is by cooling. So the first rule in keeping meat is to bring the temperature of the carcass down below 40° as rapidly as possible without freezing. Specific methods are discussed in foregoing chapters on various species.

Meat is hung or aged to allow enzymes to begin breaking down tissues, thus making the meat more tender. The trick is to hang meat long enough to tenderize it and give it good flavor, and not so long as to cause spoilage and bad flavor. The older the animal, the longer it must be aged. But recent studies indicate that some meats, venison in particular, "age" in the freezer, which makes the meat more tender. Again, chapters on different animals have recommendations for aging times.

Freezing

Except in the case of young chickens and rabbits, freezing in less than 24 to 48 hours is considered incorrect and may cause toughness. When you're cutting meat have someone help you by wrapping and labeling (always include a date) so cut meat doesn't have to lie out in the cutting room and warm up too much. Don't skimp on freezer paper. The more layers you use the better meat will keep once it is frozen.

In freezing large quantities of meat, such as a beef, if your freezer can't handle it all at once you may be able to borrow space from a neighbor or two. Pile packages loosely in the freezer so meat can freeze rapidly.

Before freezing, most meat is boned to save space and because bone and fat turn rancid more quickly than muscle. In boning a piece of meat, you remove the bone, fat, and less edible tissue from the muscle. Using a narrow-bladed knife, follow the natural divisions between muscle bundles, dissecting more than cutting, and taking the shortest, most direct route to the bone. Once you have exposed the end of a long bone, pick it up in your hand and, using your knife with the other hand, peel the muscle from the bone.

Freezing does not necessarily improve meat; the meat you eventually eat is only as good as what you start with. One exception to that rule is that trichinas, a parasite of pigs that cause the disease trichinosis in humans, are destroyed by keeping pork at 0° for 30 days.

Before the introduction of home freezing of foods, all meats had to be salted to be kept. Now, other than corned beef and tongue, pork is the only meat regularly salted. There is little use of salt pork in modern times, but a large percentage of pork that is eaten is salted and smoked, such as ham and bacon.

Dry Cures and Pickling

Pork may be salted dry or in brine (pickling) or by a combination of the two. You may buy a commercial dry sugar-cure preparation or you can try the following for 100 pounds of ham or shoulder (for 100 pounds of bacon or other thin cuts, reduce by half):

Salt, 8 pounds
Sugar, 2 pounds (brown, white, syrup, or a mixture)
Saltpeter, 2 ounces

Mix the products completely, being sure to get the saltpeter evenly distributed. Save half the mixture for resalting in a week, except for bacon, where one treatment is sufficient.

Start with meat at 40° or below and rub the salting mixture on well, leaving a ⅛-inch layer on hams and less on lean cuts. Place the salted meat in a clean crock or barrel and put it in a cold place at 36° to 40°. Hams and shoulders should be resalted in a week. Figuring on a minimum curing time of 25 days, leave bacon to cure for 1½ days per pound for each piece (larger cuts, 2 days per pound). Be sure to mark the date on a calendar. When you take the smaller cuts from the salting crock at 25 days be careful not to knock the salt off the larger cuts you leave in. Be sure the temperature is

36° to 40°. If it goes below 36° for a number of days, add that number of days to the total salting time.

For sweet pickle (brine) cure use the same weights — 8 pounds of salt, 2 pounds of sugar, 2 ounces saltpeter — and add to them 4½ gallons of drinking-quality water. Fit cold meat into a clean barrel or crock or plastic garbage can. Use enough solution to cover meat and weight to submerge it. Mark the date on your calendar. On the 7th, 14th, and 28th days remove the meat, pour off the mixture, stir, and put it all back together again. Keep it at 36° to 40° during this time.

Rock salt may be substituted for regular salt in brine mixtures. Make sure you get it completely dissolved, using hot water if necessary. If you can't get clean rock salt buy water-softener crystals. A hydrometer may be used to measure the amount of salt in the brine, but a reliable old-fashioned method is to add salt until there is enough in the mixture to float an egg.

Pickling time for hams and shoulders is 3½ to 4 days to the pound with a 28-day minimum. Light pieces of bacon need only 15 days in the cure, and medium-sized cuts, such as loin and heavy bacon, need 21 days.

If using these instructions gives you bacon too salty for your taste, try using 5½ gallons of water to the mixture instead of 4½ gallons.

If the mixture becomes sour or ropy, start over with fresh solution after scrubbing meat with hot water and a brush, and scalding and rechilling the crock or barrel.

Remove lighter cuts from brine or salt when their time is up, brushing the salt off the dry-cured cuts, and hang in a cool place until the heavier pieces are cured and ready to smoke.

Some people recommend soaking cuts in cold fresh water 15 to 30 minutes to remove excess surface salt, or you may scrub meat with hot water and dry just before smoking to make it take on a lighter color.

As you gain experience you may wish to try one of the many alternatives to the foregoing accounts of salting and brining, such as starting with dry cure for a short time and finishing with brine, pumping or injecting brine into larger hams prior to giving them a dry or brine cure, and using commercial dry and brine mixtures. Using the methods already described will be easier, however, and it will certainly be more economical than buying the commercial mixtures.

Smoking

String preserved hams and shoulders through the shank to hang in the smokehouse. Put a hardwood skewer through the end of bacon slab to square it and hang with string, using the ends of the skewer, or punch a hole near the middle of the skewered end for the string.

If you don't want the smoke flavor, meat may be hung for a week to dry and then bagged or handled as described for other meats.

Smokehouses don't need to be elaborate. They can vary all the way from a barrel and smaller hobby models to a solid brick building with steel doors. In all but the hobby size the smoke source is outside the smokehouse to prevent overheating and fire. Any tight little building, box, or barrel with a means of permitting and controlling ventilation at the top, and a roof or cover to keep out rain, may be used (see diagrams).

Smoking colors, flavors, and dries salted meats, inhibiting rancidity. Hang cured, soaked, and scrubbed pieces so they don't touch one another, and permit them to dry overnight before starting to smoke. Any hardwood, corn cobs, or hardwood sawdust may be used to produce smoke. My own favorite smoking fuel is apple and/or hickory wood. Once you get the fire going you may add green hardwood sawdust or green sticks of hardwood to cool the fire and make more smoke, although heavy smoke is not needed, just a good haze. A fire built Indian-fashion with the sticks radiating from the center like spokes will cool as it burns and won't cause problems by overheating. At any rate, don't leave a smokehouse smoking and unattended.

The first day's fire should cause a temperature of near 120° in the smokehouse to melt off excess fat from the hanging meat. After the first day, skip a day or two, then smoke for

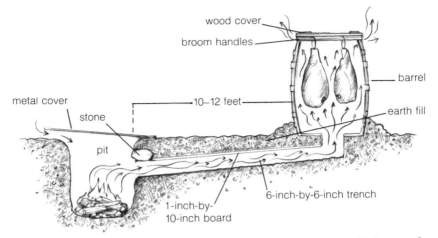

wood cover
broom handles
metal cover
stone
10–12 feet
barrel
earth fill
pit
1-inch-by-
10-inch board
6-inch-by-6-inch trench

Barrel smokehouse. Stovepipe or tile, if available, could be used for the flue.

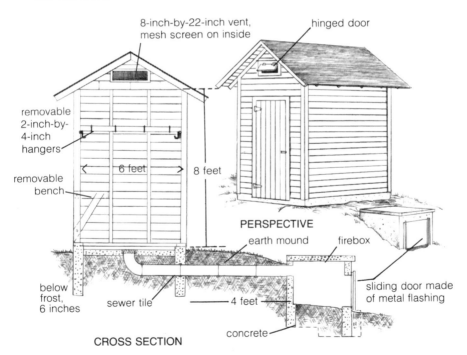

8-inch-by-22-inch vent, mesh screen on inside
hinged door
removable 2-inch-by-4-inch hangers
removable bench
6 feet
8 feet
PERSPECTIVE
earth mound
firebox
below frost, 6 inches
sewer tile
4 feet
sliding door made of metal flashing
concrete
CROSS SECTION

A frame construction smokehouse. For complete instructions on how to build different kinds of smokehouses refer to *The Canning, Freezing, Curing & Smoking of Meat, Fish, and Game* and "Build A Smokehouse, A-81." See last page of this book for information.

A homemade frame construction smokehouse.

a day at 90° to 110°. Skip a day or two again and smoke a day at only 90°. A fourth day may or may not be needed.

Cured, smoked hams and shoulders may be kept for a year or more if wrapped in heavy paper, covered with muslin bags, and hung by the bags so no insects may follow the string into the bag. Do not wrap hams or shoulders to store until a week after smoking. At that time run a sharpened wire along the bone to check for sourness. If the wire pulls out smelling sweet, with no bad odor, hams will keep all right. If not, cut the ham open and inspect it. If it smells bad, it has spoiled, and you will probably have to destroy the whole ham.

Bacon will not keep as well as shoulders and hams and should be used within a few months to prevent rancidity, even if kept in the freezer.

Cured beef tongues are sometimes smoked to add flavor, but other than brisket, cuts of beef are rarely salted and smoked in this country. One exception is jerky, which is made in many ways from lean beef, venison, and sometimes bear meat. You may make it by dry salting lean meat with a mixture of 2 pounds of salt, ½ pound of sugar, and ½ ounce of saltpeter. Pieces should be sliced no more than 1 inch thick, 2 to 5 inches wide, and 5 to 6 inches long. Pack in a wooden box at 36° with enough salt to cover them and rub salt into the meat daily, allowing moisture to drain off. After 2 weeks cut with the grain into strips no more than ⅛-inch thick. You may now dry the meat in the sun or at low heat in an oven, or smoke it lightly until the meat becomes hard and dry. Recipes for jerky are as numerous as people who make it. You can vary your salt with spices such as black pepper and ginger, or condiments such as soy sauce mixed with ginger (teriyaki). When properly made, jerky will keep without refrigeration for months.

Norwegians and other Scandinavians salt and smoke reindeer, lamb, and goat with a method that works well on venison, leaving it hard on the outside and able to be stored for months, while the insides remain tender enough to be cut into thin slices similar to chipped beef. Although I have not tried it, I see no reason why the method would not work on beef as long as it is lean with all fat removed. I have tasted the lamb, goat, and reindeer in Norway and have processed and eaten the venison at home, using the recipe translated by a friend in Bergen. The only problem with the venison is that people like it so well that once you start serving it, it goes too fast.

Trim the "ham" of a deer or leg of lamb or goat so you have only the part from the ball of the ball-and-socket joint to the hock. Put it in a wooden box that has a few tiny holes or cracks to allow moisture to drain out, and rub it two or three times a day with the following mixture:

   3½ pounds salt
   1⅓ ounce saltpeter
   1⅔ pounds sugar
   3½ ounces Lyle's Golden Syrup (partially inverted refiner's
      syrup) or Karo syrup or honey
   7 ounces water

Turn the meat each day and remove at the end of a week. Thereafter put the meat in the following brine for a week at 36° to 40°:

42 quarts of water
26 pounds of rock salt
4½ pounds of sugar
5 ounces of saltpeter

Use hot water to dissolve the rock salt, but cool the mixture down to 40° before putting the meat in. It takes a deep crock to allow you to cover a leg of venison completely, so I use a plastic garbage can that holds 48 quarts (12 gallons). Stir the mixture every other day or so and don't be upset if some of the salt never does dissolve. At first you will have to weight the leg to hold it down. After a few days it will stay under in the solution by itself.

After a week remove the leg and let it hang for 5 to 6 days to dry before smoking. After smoking, hang uncovered in a cool place with good air circulation for 2 to 3 months (the Norwegians recommend 4 months). By that time it will be hard as stone and even look like stone. However, it may be sliced easily with a sharp knife, the thinner the better. The Norwegians serve it with cinnamon-spiced apple juice and their porridge in a traditional meal. Americans seem to like it with crackers, cheese, and beer.

Corned Beef and Tongue

Don't dump the brine out after salting the leg of venison or lamb. It should still be good to use for beef tongue, and small, 2 to 3 pound pieces of lean beef or brisket. The small pieces will cure well in the mixture in 4 to 7 days (3 to 4 days for lamb tongue), and the larger brisket will take up to 2 weeks.

The brisket, of course, is used for corned beef. Like the corned tongue it may be eaten as is, stored in the refrigerator for a few days, or kept frozen up to 4 months. Brisket and tongue are salted more for flavor than to preserve them, so you may vary the time in the brine to suit your taste, allowing less time for less salty meat. If you want to corn just beef or tongue you may start with only a quarter or a half the amount of mixture needed for a leg of venison or lamb.

Corned tongue is sometimes smoked to add flavor, and as I said previously, I can see no reason why tongue or other cured lean beef could not be smoked in the same way as the venison or lamb to make a product like chipped beef.

Pickled pigs' feet may be made with feet properly cleaned and prepared by removing the toes and dewclaws at butchering and removing all hair, dirt, and glandular tissue between the toes as soon as they are properly chilled. Put into a brine using 4½ quarts water, 1 pound salt, ¼ pound sugar, and ¼ ounce saltpeter. If the cure is kept at 36° to 40°, the feet should be in the brine for 2 to 3 weeks. Weight them to keep them from floating. After curing, simmer the feet in fresh water until they are tender, then chill and pack in vinegar with bay leaves, cloves, and allspice. You may use them at once or store them in the vinegar for 3 weeks.

Souse is made by cooking cured or uncured feet and pork trimmings in a little water until meat slips off the bones. Season the meat and strained stock with vinegar and spices, bring to a boil, put in molds, and allow to jell.

Sausage

There are more recipes for sausage than there are butchers. By far the most popular sausage is fresh pork sausage. It is made using pork trimmings, ⅓ fat and ⅔ lean. Cut the trimmings into pieces small enough to be easily put in the sausage grinder, spread the seasoning mixture over the meat, and grind. You may put the seasoned ground meat directly into sausage casing and use or freeze it, or by adding a little water to make it soft and sticky, you can pack the sausage in pans and refrigerate before slicing and frying as patties.

Following is one recipe for making sausage out of 50 pounds of pork:

¾ cup salt
¾ cup black pepper
⅓ cup sage
1 tablespoon coarse red pepper
1 tablespoon cayenne pepper

Mix spices well before sprinkling over meat as described above.

You can vary the above to suit your taste by changing the proportions or by adding ¼ to ½ ounce of ground cloves, ½ ounce of nutmeg, and 6 ounces of sugar. Don't use sugar in sausage that is to be stored for long.

Ground pork will keep longer in the freezer if you don't salt it or spice it until it is thawed and ready to use. For 2 pounds of thawed ground meat you may add and mix by hand:

2½ teaspoons salt
2 teaspoons ground sage
1 teaspoon pepper
¼ teaspoon ground cloves (or ½ teaspoon nutmeg)
½ teaspoon sugar

To make sausage to be smoked, increase the salt for 50 pounds to ⅞ cup, instead of ¾ cup. Stuff in casing (add water if too stiff) and store sausage for 24 hours in a cool (36° to 40°) place. Smoke and dry at 70° to 90° for a day or two until the sausage takes on a dark mahogany color. This sausage will not keep in warm weather and so should be eaten soon after it is made.

To make "hot" Italian sausage from 50 pounds of meat use:

¾ cup salt
¾ cup red pepper (cayenne)
¾ cup paprika
¾ cup black pepper
¾ cup coarse red pepper
⅓ jar of fennel seed (one ⅜-ounce jar)

To make "sweet" Italian sausage from 25 pounds of pork trimmings use:

7 tablespoons salt
7 tablespoons pepper
5 tablespoons fennel seed
1½ tablespoons sugar

To make kielbasa (Polish) sausage from 20 pounds of pork trimmings use:

6 tablespoons salt
6 tablespoons black pepper
1 tablespoon sugar
2 buds fresh garlic

Some people use lean beef from a "shelly" old cow, mixed 50/50 with pork, to make Italian or Polish-style sausage.

On all of the above follow the directions listed for mixing and grinding fresh pork sausage. Stuff the sausage in casings directly from the grinder if possible, or use a sausage stuffer. Animal casings are in packed salt when purchased. Rinse in fresh water just before using. Slide the casing up on the spout of the stuffer accordionstyle, hold the end with two fingers, fill a little, and tie where your fingers were. Now, guiding the casing with your fingers so it doesn't come off too fast, stuff slowly, and tie off sections that loop into rings about 8 inches in diameter.

Commercially made kielbasa is cooked before being refrigerated and sent to market. For home use it may be put directly into the freezer. Other sausages are put directly into the freezer, but since they contain salt should be used within 3 to 4 months.

Once you have mastered making basic sausage you will want to try other kinds or vary the foregoing by using lean venison and pork fat, or by using different proportions or spices. You may also wish to try bologna, headcheese, liver sausage, and scrapple. Following are basic recipes for these foods with simplified directions which you may wish to alter.

BOLOGNA SAUSAGE

60 pounds of beef
2 pounds salt
40 pounds pork trimmings
1 ounce saltpeter
3 ounces black pepper
1¼ ounces coriander
1 ounce mace
10 quarts cold water
Onions if desired, ground fine

Grind the chilled beef with 19 ounces of salt, using a coarse grinding plate, and cure in a cold place for 48 hours. The next night grind 13 ounces of salt with the pork and cure overnight. Grind the beef again to ⅛ inch, then add pork and regrind, adding the saltpeter, onions, and spices as you do. Add water and mix until the mass is sticky.

Stuff into beef casings or muslin bags and allow to cure in a cool place overnight. Smoke at 110° to 120° with plenty of ventilation for about 2 hours.

Put immediately into 170° water and cook at that temperature until the casings "squeak" when pressure of the thumb and forefinger is suddenly released. This will take 15 to 90 minutes depending on the size of the casings. Chill in cool water and hang in cool place or freeze.

SCRAPPLE

Start out as with headcheese, but after boning grind everything through the fine plate, return to the strained cooking water ("soup") and bring to a boil again. A cereal mixture of 7 parts cornmeal and 3 parts buckwheat flour is one example of what is added to the mix in a proportion of 1 part cereal to 3 parts of meat and soup. To mix evenly add some cool soup to the cereal mixture first to smooth out lumps.

Boil for 30 minutes, stirring constantly. Add the following seasoning toward the end of the boiling for 50 pounds of mix:

1 pound salt
1½ ounces black pepper
1½ ounces sweet marjoram
1½ ounces sage
½ ounce of nutmeg
¼ ounce mace
1 pound ground onions (if desired)
½ ounce red pepper

Pour into shallow pans and chill. You should be able to slice the scrapple when it's cool and fry without its crumbling. If you can't, vary your method by using different proportions of cereal to meat mix. Dipping the cold scrapple in beaten egg before frying may also help to smooth it out.

LIVER SAUSAGE

This is 10 to 20 percent liver made into a sausage with the same pieces mentioned above for headcheese. Cook as above without the liver. Bone the meat and add the livers, which have been scored with a knife, and cook for 10 minutes more. Grind all cooked material moderately fine and add about ⅕ as much of the cooking water by weight, making a soft mixture. Add the following to 50 pounds and mix well:

1 pound salt
2 ounces black pepper
1 ounce sage
¼ ounce red pepper
1 ounce allspice

Stuff into beef casings and simmer in water until it floats. Then chill in cold water and hang to drain in a cool place.

HEADCHEESE

Cook head, tongue, skin, hearts, snouts, even ears and other pieces of pork in water until the meat slips from the bone. Remove bone and cartilage, and grind or cut everything into ½-inch pieces, except skin, which should be ground fine. Mix enough of the water in which the meat was cooked to the fine- and coarse-cut pieces to make them soft but not sloppy. For 50 pounds of the mix add the following to taste and bring to a boil:

1 pound salt
2 ounces black pepper
¼ ounce red pepper
½ ounce ground cloves
½ ounce coriander
1 ounce sweet marjoram

Pour into shallow pans and chill. The headcheese should jell so you can slice it. Served with homemade bread and butter and home fries it is an unusual treat.

Other Preserved Meats

Meat may be canned at home, but canning should only be done by steam pressure. Water-bath processing and other means are not hot enough, a temperature of 240° being needed for sterilization. The taste of home-canned stews or meat and gravy is something that is hard to duplicate, but the difficulty in making them has nearly eliminated their use in favor of other more modern or easier methods of meat preservation such as freezing.

Lard can be preserved without canning. To "render" lard means simply to heat it until most of the moisture is boiled off. This is accomplished with pork fat by grinding or cutting it into small pieces and heating. The temperature will go from 212° (boiling water) to 255° when sufficient moisture is gone for lard to keep. When the solid matter (cracklings) turns brown and starts to sink the lard is ready. Strain it through cheesecloth into 10-pound containers, chill, and cover. Once chilled it should keep for months in a cool place, or longer if frozen. The cracklings are edible and considered a delicacy by Southerners.

12•RECIPES

EVERY COOK KNOWS HOW to prepare steaks, chops, roasts, cutlets, broilers, and fryers, the cuts generally purchased in the supermarket or butcher shop. But when a whole animal has been slaughtered, cut up, and put in the freezer there is a challenge to use the more unusual cuts in a creative way. Cooking the small game bagged by the hunter of the family is certainly a challenge, too.

Most good cookbooks have recipes for cooking breast, shoulder, neck, etc., and the sundries such as sweetbreads, brains, and kidneys. There are also cookbooks specifically dealing with wild game.

The following recipes have been gathered from many good cooks among friends, family, professional chefs, and members of a conservation club. The emphasis here is on the use of more unusual meats, including a few recipes for especially delicious ways of preparing those cuts that are eaten more often. The recipes are the creations of these cooks, and like most recipes they can be varied as each cook adds a personal touch. Enjoy them as they are presented, then perhaps alter them a bit to make them your own.

SHORT RIBS OF BEEF

The main thing about shortribs is that they have to be cooked until they are tender. You can bake or pressure cook them, but the best way seems to be cooking slowly in a crock pot. In fact, any recipes using stew meat or other cuts needing tenderizing as they cook can be adapted to crock-pot cooking.

3 pounds of beef shortribs
1 teaspoon salt
½ teaspoon pepper
1 cup flour
cooking oil or beef drippings
2 medium onions sliced
1 cup carrot chunks
1 cup water
2 tablespoons catsup
1 tablespoon vinegar
¼ teaspoon thyme

Roll shortribs in seasoned flour and brown well in large skillet in beef drippings or oil. Put carrots in the bottom of the crock pot, add browned ribs and onions. Add water and seasonings. Cover and cook on low heat for 7 to 10 hours or on high 4 to 6 hours. Remove meat to platter or bowl and thicken gravy with flour-water mixture. Most of the meat will have fallen from the bones, and you can trim away any bone and fat before serving in the gravy.

STEAK-AND-KIDNEY PIE

If you've ever traveled in the British Isles you've probably enjoyed a steak-and-kidney pie on a pub lunch. The following version can just as well be made with stew meat instead of round steak.

Just a word about rib roast, or any other oven roast of beef: the British do it best. Heat oven to 475° and roast at that temperature for 15 minutes. Turn oven down to 425°. Beef should be cooked only 15 minutes per pound. (Those first sizzling 15 minutes may set your smoke alarm off.) Yorkshire pudding is a great addition to any British-style meat dish.

1 pound stew beef or round steak
1 beef kidney, separated into 1-inch pieces
   (peeled and fat removed)
seasoned flour for shaking meat
2–3 tablespoons hot oil or beef drippings
1 small onion, chopped (optional)
1 tablespoon Worcestershire sauce
2 cups water with beef bouillon cube (or 2 cups consommé)

Shake pieces of meat in flour in brown paper bag and brown them in oil or drippings. Cook in pressure cooker for 15 minutes at 15 pounds pressure, or give the meat 1½ to 2 hours regular cooking, adding liquid as necessary. Thicken with 2 tablespoons flour mixed with ½ cup cold water and simmer a few minutes. Pour into baking dish and top with baking-powder-biscuit dough (Bisquick may be used) rolled to fit. Bake at 450° until brown.

SPICY UPSIDE-DOWN HAMBURGER PIE

Recipes using hamburger are endless in variety. A favorite of our family when the children were young was called "Sophie Pie," having come from the wife of our country doctor. The following is a spiced-up version created by one of those children now grown up. Use this, therefore, for those with grown-up tastes, and simply omit anything too spicy for the little ones, thus going back to Sophie's recipe.

1 pound ground beef
½ cup chopped onions
1 can condensed tomato soup
1 small can chopped green chilis
1 tablespoon Worcestershire sauce
1 tablespoon Dijon mustard (or Outerbridge's
   Hot Mustard Sauce)
½ teaspoon fresh ground pepper
½ teaspoon Tabasco
½ pound sharp Cheddar cheese, in thin slices
Bisquick and required milk, or homemade biscuit dough

Brown ground beef and onions in skillet (preferably a cast-iron one so it can be placed in the oven). Drain meat. Add all but last two ingredients (cheese and dough) and heat through. Prepare biscuit and roll into a piece large enough to cover skillet. If skillet cannot be placed in oven, transfer meat mixture to similar-size shallow baking dish and top with sliced cheese, then rolled biscuit dough. Bake at 350° for 20 minutes or until browned. Remove from oven and let set a few minutes, then place large plate over top of skillet or dish and flip so pie is served meat side up.

BEEF OR VENISON BURGUNDY
(and Meat Pie)

Marinate 1½ pounds of stew meat overnight in red wine to cover, in refrigerator. Sauté ¼ pound sliced mushrooms in 2 tablespoons butter or margarine. Lightly flour meat (by shaking in paper bag), and brown, with mushrooms, adding more butter or margarine as needed. Add following ingredients:

1 cup chopped (in chunks) carrots
1 cup Burgundy wine (or ¾ cup wine and ¼ cup water)
1 tablespoon tomato paste (or catsup)
1 beef bouillon cube
1 bay leaf, crumbled
¼ teaspoon thyme
¼ teaspoon salt
¼ teaspoon pepper

Cook in pressure cooker at 15 pounds pressure for 12 to 15 minutes, or allow 1½ to 2 hours regular cooking (adding more liquid if needed). Thicken with flour and water shaken together to make thin paste. Add sprinkling of chopped parsley.

This may be put in casserole and topped with pastry or biscuit dough rolled to fit (mix may be used, such as Bisquick), and then put in 450° oven until browned. Serves 4 to 6. Double for a crowd. Without the crust it's nice served on a buffet.

SAUERBRATEN
(Beef or Venison)

Beef, round or rump, or any venison roast (3 to 4 pounds)
1 pint vinegar
3 bay leaves
12 peppercorns
6 whole cloves
1 sprig chopped parsley
2 tablespoons seasoned flour
¼ cup butter or beef drippings
1 cup sliced onions
1½ cups carrots, cut up
1 dozen gingersnaps
1 tablespoon sugar

Wipe meat dry. Place in earthenware crock, add vinegar and enough water to cover. Add bay leaves, peppercorns, whole cloves, and parsley. Cover and place in refrigerator. Leave 3 to 4 days, turning the meat daily. Drain off liquid, reserving it for sauce. Rub meat on all sides with seasoned flour. Brown it thoroughly in butter or drippings; add onions, carrots, and 2 cups spiced liquid. Cover and simmer gently about 2 hours or until tender. Remove to hot platter. Add gingersnaps, rolled into fine crumbs, and sugar to gravy. Cook about 10 minutes. Add more salt if desired. Makes 6 to 8 servings. Serve with potato dumplings or noodles and red cabbage.

CHINESE SUB GUM

David Silvernail raises three hogs every year to feed his family of six. The following are two of the family's favorite recipes. David's wife, Irene, cooks them using their home-butchered pork and home-made sausage.

4 tablespoons margarine
2 pounds lean pork, thinly sliced
1 large onion
6 stalks celery, sliced
Salt and pepper to taste
2 cups hot water
8-ounce can water chestnuts (sliced)
2 cans sliced mushrooms (drained)
2 tablespoons cornstarch
2 tablespoons cold water
2 tablespoons soy sauce

Melt margarine in a heavy skillet, brown the pork and onion, add celery, and salt and pepper. Cook 5 minutes or so, add the hot water, water chestnuts, and mushrooms, bring to a boil and simmer 10 minutes. Combine cornstarch, water, and soy sauce, stir a little at a time into the meat mixture, and cook, stirring constantly, for about 5 minutes or until thickened. Serve on cooked rice.

SAUSAGE CHEESE CASSEROLE

1 pound pork sausage
½ cup chopped onion
1 cup chopped green pepper
1 large can tomatoes
6-ounce can tomato paste
2 cups water
½ teaspoon salt
¼ teaspoon oregano
¼ teaspoon black pepper
8 ounces elbow macaroni, cooked
Thin slices Cheddar cheese, grated Parmesan cheese

Cook pork sausage, browning well (drain off some of the grease if sausage is a bit fatty), add chopped onion and green pepper, and sauté until soft. Add tomatoes, tomato paste, and water. Mix well. Add salt, oregano, and black pepper. Heat to boiling. Add cooked macaroni and stir well. Transfer to buttered casserole dish, top with slices of Cheddar cheese, then sprinkle top with Parmesan. Bake in 350° oven 30 to 35 minutes until bubbly. Serves 6 to 8.

LINDA'S LEFTOVER HAM CASSEROLE

Smoked ham is generally baked (and ham steaks broiled) and varied by different glazes: fruit, brown sugar and mustard, etc. It is the leftover ham that offers a challenge for use in many tasty recipes, usually casseroles, and, of course, the bone is the basis of good thick pea soup, or of a bean or lentil soup. This casserole is enjoyed by the youngsters of the family.

2 cups cooked ham, cut up
8 ounces noodles, cooked

Cook to make sauce:
¼ cup butter
¼ cup flour
1½ cups milk, or 1 cup milk, ½ cup light cream
½ cup chicken broth
½ teaspoon lemon juice
1 teaspoon prepared mustard
pepper

When thickened remove from heat and add:
½ cup mayonnaise
pinch rosemary
2 teaspoons parsley

In casserole dish layer noodles, meat, sauce, and repeat. Bake ½ hour at 350°, uncovered.

WIENER SCHNITZEL

The ordinary way to prepare veal cutlet is simply to bread it, sauté, and serve with tomato sauce. But once you try pounding it thin and making veal marsala (with mushrooms and marsala wine), veal parmigiana (with tomato sauce and melted mozzarella cheese), and other variations such as veal scallopini (with white wine), you'll never go back to the old way. One of the best ways to eat your veal cutlets is in the German style as Wiener schnitzel.

1¼ pound veal cutlets, pounded thin
1½ cups flour
2 eggs plus 2 tablespoons water, lightly beaten
2½ cups fresh bread crumbs
2 tablespoons cooking oil
6 tablespoons butter
1 fresh lemon
¼ cup capers

Dip veal in flour, then egg, then bread crumbs, patting to coat well. Place in no more than 2 layers on plate, cover with wax paper, and refrigerate for at least ½ hour. Heat oil in skillet. Cook veal quickly over medium-high heat until coating is browned. In separate pan melt butter. Serve veal with melted butter, lemon wedges, and capers. Serves 4.

GESCHNETZELTES ZURCHERART
(Diced Veal Zurich Style)

This delicious veal dish is Swiss in origin and is interpreted by two cooks, one the owner-chef of the Swiss Hutte, a fine restaurant in the Berkshires at the foot of Catamount Ski Area near Hillsdale, New York; the other Annelies Oswald, a young Swiss woman who is our family's "Swiss daughter," having lived with us many years ago through an international exchange program. Annelies specifies using stew veal, although Tom Breen of the Swiss Hutte uses the trimming of cutlets after cutting what he uses for Wiener schnitzel.

10 ounces fresh mushrooms
Juice of ½ lemon
2 ounces margarine or butter
1 small onion, finely chopped
1 pound 2 ounces stew veal, cut thumbnail size

Sauce:
3 ounces white wine
3–5 ounces mushroom juice and bouillon (stock)
4–5 ounces cream

Seasoning:
Red pepper and/or curry (to taste)
1 teaspoon salt
Black or white pepper, freshly ground

Slice mushrooms, cook slowly for 2 minutes with lemon juice in small pan. Season the mushrooms with salt and pepper and set aside the juice for sauce. Melt butter in skillet, add onion, and sauté until golden, then add veal. Cook over quite high heat, turning constantly. After about 2 minutes it should be almost white. Set meat aside and add wine to same skillet, cooking a little, add mushroom juice and stock, cook until sauce thickens slightly. Add cream. Season, sprinkling the pepper on the meat before adding veal and mushrooms to sauce to be heated quickly. Serve with *Rösti* (potatoes fried Swiss style, see below) or rice.

GESCHNETZELTES KALBFLEISCH SWISS HUTTE

24 ounces veal (trimmings from cutlets; pork may also be used)
Salt and pepper to taste
1 tablespoon oil
2 tablespoons butter
4 tablespoons shallots, chopped fine
2 cups sliced mushrooms
4 ounces white wine (dry)
1 pinch thyme
8 ounces chicken or veal stock (unsalted)
8 ounces heavy cream
1 teaspoon cornstarch

Cut veal into 1½-inch by ½-inch pieces, season with salt
and pepper. Put oil and 1 tablespoon of butter in saucepan,
brown the meat, and remove to warm platter. Don't wash
pan; add other tablespoon of butter to pan, add shallots and
mushrooms. Cook 1 minute. Add veal, white wine, and thyme.
Reduce wine by simmering until wine has almost evaporated.
Add stock and reduce by half. Add cream mixed with
cornstarch and cook to sauce consistency. Correct seasoning
if necessary. Serve with *Rösti* potatoes.

RÖSTI

Cook potatoes in their jackets and peel at once. Cut in fine
slices or grate. Heat plenty of butter or margarine, add
potatoes and a little salt, fry on a hot flame, turning all the
time, until they are golden brown. Lower flame, press the
potatoes down, and leave them for another few minutes until
a golden crust forms underneath. Turn out so that the crust
is on top. (A finely chopped onion may be added while cooking.)

HIGH VALLEY LAMB CURRY

The lamb raised at High Valley Farm is perhaps the most delicious in the world, so of course their broiled chops and roast leg of lamb are wonderful. This recipe was concocted to use stew lamb in an interesting way. Most important, be sure the meat is completely free of fat and bone before it is frozen.

Lightly flour 1½ to 2 pounds stew lamb, trimmed and cubed. Brown in 2 tablespoons cooking oil in heavy kettle or pressure cooker. Add, in order:

1 onion, chopped
2 crushed bay leaves
1 tablespoon garlic wine vinegar
1 tablespoon curry powder
¼ teaspoon dried thyme
1 teaspoon salt
1 can condensed consommé, undiluted
2 tablespoons catsup
½ cup carrots, cut in chunks
½ cup (or 4-ounce can) mushrooms

Pressure cook at 15 pounds for 12 minutes, or allow 1½ to 2 hours regular cooking. Thicken with 2 tablespoons cornstarch mixed in ¼ cup water. Serve over fluffy rice.

ROAST WHOLE LAMB

Martin Stosiek raises his own lambs, and every year he roasts a whole lamb on a spit over a wood fire, which takes a long time and a lot of patience — plus several friends to keep turning it! For lamb ribs, a little-used cut, he provides a Somali recipe adapted by his wife, Gertraud.

First prepare the place of fire by building crossbars, two at each side a foot apart, and setting them up about 8 feet apart. They should be strong enough to support the lamb and stable enough to withstand the turning of the spit. Attach crosspieces about 10 inches apart to the two posts at each side in order to raise and lower the level of the lamb over the fire. (These will resemble the uprights for a bar-rail gate.) Cut a straight 12-foot hardwood pole for the spit, 2 inches thick on the small end, 4 inches on butt end. Debark and sand pole. Drill hole in butt end to insert turning handle.

Collect wood for fire, thin branches for starting, thicker to build up fire. Split logs to build embers, using some softwood, mostly hardwood. Recruit people for turning spit. Start fire in morning, being aware of wind. Build up a good bed of embers, add new wood from side away from wind and push embers under where lamb will be. Test heat by placing hand over the fire (at the same distance that the lamb will be from the fire): if you have to remove it on the count of one the embers are very hot, by the count of three you have medium heat, and by five slow heat. You should have medium heat.

To prepare the lamb, leave head on and don't cut pelvis. Insert spit through lamb lengthwise, stretch out legs, and tie to spit. Secure lamb so that it can turn only with the spit. Brush meat with cooking oil and insert six cloves of garlic, one in each leg and one in the loin on each side. Place spit over medium-heat embers and start turning, slowly but steadily for 3 to 4 hours or desired doneness. The turning self-bastes the lamb (but add liquid to the turner of the spit). Serve with potatoes and foil-wrapped corn baked in the embers.

SWEET-SOUR KIDNEY STEW

Kidneys can be used in many ways; the more tender lamb, veal, or pork kidneys may be broiled or they may be sautéed with wine, French style. Lamb kidneys can be broiled along with lamb chops, sausages, tomatoes, and mushrooms (and lamb liver if you wish) for a fine mixed grill.

Beef kidney, however, is better braised or made into stew, as in this German-style recipe handed down through a friend's family.

2 beef kidneys
2 tablespoons butter or margarine
1 onion, chopped
4 cloves and a crushed bay leaf made into bouquet in cheesecloth
⅓ cup sugar
⅓ cup vinegar
flour for gravy

Soak kidneys in lightly salted water for 1 hour in covered pot. Drain and rinse well. Cut kidneys into squares, making sure to remove all fat. Melt butter or margarine in pot and sauté chopped onion. Add kidneys and bouquet, and cover with water to about 2 inches above kidneys. Cook 1½ hours. Slowly add sugar and vinegar, cooking slowly ½ hour. Taste and add more sugar or vinegar if the kidneys have become too sweet or sour while cooking. Mix flour and water to add to thicken gravy. Serve with dumplings, potatoes, or noodles.

LAMB RIBS WITH ROSE WATER

4 breasts of lamb, sliced into spareribs

Mix and pour over meat, marinate overnight, drain.
Juice of 3 lemons
Small can of frozen orange juice
2 tablespoons rose water
½ teaspoon powdered cloves
1 teaspoon powdered ginger

Mix:
2 teaspoons salt
½ teaspoon cayenne pepper
½ cup chopped parsley
½ teaspoon pepper
½ cup golden raisins

Place marinated meat in floured pan, sprinkle with the above dry mixture, place in cold oven, then turn oven to 350° and bake for 1 hour. Sprinkle with raisins. Pan drippings make good gravy. Serve with rice and fried bananas.

STUFFED HEART
(Beef or Venison)

A good way to use the heart from your beef or venison is the slow cooking of a crock pot. A tasty stuffing will make it a more interesting dish.

1 beef or venison heart, slit open lengthwise
Mushroom stuffing:
   10 slices bacon, diced
   1 medium onion, finely chopped
   ½ pound mushrooms, sliced (or two 4-ounce cans, drained)
1 garlic clove, minced (optional)
½ cup oil-and-vinegar salad dressing
1 cup beef broth (bouillon or consommé)

Trim fat and remove tubes from heart and wash it well in salt water. Pat dry. To make stuffing, sauté bacon, onion, and mushrooms in skillet until onion is soft. Fill heart with stuffing, drained. Skewer or sew with string to fasten heart closed. Place in crock pot. Add garlic clove, salad dressing, and beef broth. Cover and cook on high for 1 hour, then on low for 7 to 9 hours. Thicken gravy with a little flour before serving.

SWEETBREADS AND LOBSTER POULETTE SAUCE

Sweetbreads (and brains) should be soaked in cold water for 15 minutes and simmered with a tablespoon of lemon juice or vinegar to keep them white. Membranes should be removed. After precooking they may be prepared very simply by dipping in egg and crumbs and sautéing or broiling (merely brushed with melted butter). They may also be served in a sauce or in salad with chicken. To be a bit more adventurous, try this recipe created by Tom Breen of the Swiss Hutte, using some simple ingredients such as canned soups. It will establish your reputation as a gourmet cook, and no one need know that it's simple to make. (Sweetbreads may also be served with chicken in a sauce, or with shrimp, as noted. In the latter case the seafood soup used might be a canned shrimp soup.)

2 pairs sweetbreads
1 tablespoon lemon juice
1 teaspon salt
1½ cups fresh, frozen, or canned lobster meat (or shrimp)
⅓ cup butter
3 egg yolks
1 tablespoon cornstarch
1 tablespoon lemon juice
1 can (10¼ ounces) oyster stew, condensed
1 can (10½ ounces) chicken broth, condensed

Cover sweetbreads with water. Add lemon juice and salt, simmer 10 minutes. Drop sweetbreads into cold water. Remove membranes and slice sweetbreads. Sauté sweetbreads and lobster (or shrimp) in hot butter for 2 minutes. Keep warm. In another pan combine egg yolks, cornstarch, and lemon juice. Stir in oyster stew and chicken broth. Cook over low heat, stirring constantly until sauce thickens. Pour over sweetbreads and lobster or shrimp, reheat (do not allow to boil). Serves 4.

VENISON PATÉ

When our daughter Meg's husband became a deer hunter and shot his first deer he wasn't sure he liked venison, until she created this gourmet paté. (Meg is one of the cooks who contributed several recipes for this book.)

2 tablespoons butter
¼ cup flour
⅔ cup milk
¼ cup chopped onion
¼ cup heavy cream
¾ pound bacon
1 pound deer liver
2 eggs, lightly beaten
¼ cup applejack or brandy
2 teaspoons salt
½ teaspoon fresh ground pepper
½ teaspoon ground mustard
½ teaspoon allspice
½ cup chopped cornichons or gherkins
½ teaspoon thyme

Melt butter. Stir in flour. Cook 1 minute while stirring. Stir in milk to make thick sauce. Remove from heat. Add onions and cream.

Cook 5 slices of bacon crisp. Crumble and add to sauce.

Chop liver fine in food processor using metal blade or put through meat grinder. Add to sauce. Stir in all other ingredients, mixing thoroughly. Press into greased shallow oven-proof dish.

Partially cook remaining bacon. Lay across top of paté.

Place dish in bain-marie. (Pan filled with enough water to come almost to the top of the dish.) Cook in 350° oven for 1½ hours or until liver is brown and any liquid is clear. Let stand until cool. Paté can be turned out or served from dish. Serve on lightly toasted slices of French bread with sliced cornichons or gherkins.

VENISON STROGANOFF

Many of our quite sophisticated country cooks use a recipe like this to prepare some of their hunter husbands' venison for a party buffet.

2 tablespoons cooking oil
⅔ cup chopped onion
1 clove of garlic, chopped
2 pounds venison, cut in 1-inch cubes*
2 tablespoons butter
8 ounces sliced fresh mushrooms
1½ cups sour cream
two 8-ounce cans tomato sauce
1 teaspoon salt
¼ teaspoon fresh ground pepper
1 tablespoon Worcestershire sauce
2 tablespoons flour
2 tablespoons water

Heat cooking oil in large, heavy frying pan over medium heat. Add onion and garlic and cook till onion is yellow. Add venison and cook until browned. In another skillet, melt butter over medium-high heat. Add mushrooms and cook until they have absorbed the butter and just start to give it up. Add to venison. Stir in all other ingredients except flour and water. Reduce heat to low and cook 2 hours. Stir together flour and water. Stir into venison and let cook 10 minutes, or until sauce is thickened.

*Use one of the tender cuts, such as round. If you're using stew meat, marinate for several hours in red wine with chopped onion and crushed juniper berries. The marinade can be strained and added to the stroganoff.

VENISON SAUTÉ

Here is a fine way to serve venison, truly a dressed-up company dinner. Dinner plates should always be warmed, but this is particularly important when serving venison, and also lamb. They must be eaten good and hot, and not cool on the plate.

1 pound venison cutlets or boneless loin chops
1 cup red wine
¼ cup chopped onion
6 juniper berries, crushed
6 tablespoons butter
¼ cup red-currant or apple jelly
¼ cup heavy cream (optional)

Marinate venison several hours in wine, onions, and juniper berries. Remove meat, strain marinade, and save ½ cup. Melt butter over medium-high heat. Sauté venison until just pink in the center. Remove to warm plate and keep warm. Add reserved marinade and jelly to pan to deglaze. Stir to loosen any bits stuck to pan and to help melt jelly. Cook about 5 minutes over medium heat, stirring occasionally until reduced somewhat and thickened. Pour over meat and serve. For a richer sauce, after wine and jelly have thickened, add cream and cook about 2 minutes more.

SHERIFF PROPER'S CAMP CUTLETS

Our county sheriff, a neighbor and a deer hunter like most of the men hereabouts, enjoys being camp cook as much as he enjoys hunting, and we like his simple recipe for a quick dinner. Other friends, incidentally, use the same recipe to make a venison snack or hors d'oeuvre.

Cut tender venison in very thin small slices (scallops); then just dip in egg/milk mixture and Italian bread crumbs. Fry quickly in hot oil and serve with spaghetti and sauce.

RACCOON OR SQUIRREL

The kids in the Conservation Club at the Roe Jan School in Hillsdale, New York, feast on all kinds of wild game they bring to meetings. Their adviser, Jim Colclough, suggests the following method for easy and tasty cooking.

The key to eating this wild game is to pressure cook it at 10 pounds for 10 minutes. Then separate it from the bones. (It should first have been cut into loins, legs, backs, etc.) Dip meat in an egg-and-milk mixture and then shake up in crushed cracker crumbs and fry in butter.

"It doesn't taste like chicken, the way people say," Jim says. "It tastes like raccoon and squirrel."

He also cooks small chunks of the meats and uses them in spaghetti sauce.

VENISON ROLADEN

Everyone seems to have their favorite recipes for venison. While Jim, a science teacher, hunts and does home butchering and smoking, his wife, Lynn, a home-economics teacher, comes up with wonderful recipes enjoyed even by those most convinced they don't like venison or other game. Here are a couple of her specialties.

Cut loin of venison into thin strips, 1 by 7 inches. Sprinkle with parsley, salt, pepper, and minced onion. Lay a slice of bacon on it and roll up, securing with toothpick. Cook over charcoal.

SPEEDIES

This recipe originated in New York State's Southern Tier and is equally adaptable to venison, raccoon, or beef. Chunks of stew meat are marinated 8 to 24 hours per pound.

Marinade:
⅓ **cup oil**
⅓ **cup vinegar**
2 teaspoons garlic powder
1 teaspoon parsley
1 teaspoon rosemary
1 teaspoon garlic salt
1 teaspoon onion powder
½ **teaspoon pepper**
1 teaspoon sweet basil
1 teaspoon celery salt
1 bay leaf

After marinating, barbecue meat on skewers.

MONKEY ON A STICK

A similar recipe originates in Hawaii. Cut venison (or beef round steak) in very thin narrow strips with the grain. Thread on individual bamboo skewers and marinate in soy sauce. Sprinkle with ginger and cook quickly over fire coals. These make an unusual hors d'oeuvre.

WILD ANIMAL WHATEVER

Carol Lee Miller has fed her large, growing family whatever her husband, Jake, and her sons have brought home from hunting — hence the names she gives her recipes, which she says apply to raccoon, woodchuck, and rabbit, and which follow in Carol's own words.

Cut up animal as you would a chicken. If not sure of its age, parboil in water seasoned with salt, pepper, onion, celery, basil, and garlic. Then melt butter in a skillet and brown the meat, seasoning with salt, pepper, garlic powder, celery salt or powder, onion powder, basil or thyme, and parsley. Don't be afraid to use lots of seasonings.

After browning, smother meat in onions, celery, and mushrooms (if handy). Add water and let simmer until tender, adding more water as it cooks down. With the water add any cream soup you have on hand. If no soup is available and you want a creamy sauce, add milk with 1 tablespoon vinegar, plus a little flour for thickening. Add a little Worcestershire sauce for pizazz if you'd like. Be creative! Sprinkle parsley flakes over your creation just before serving, and serve with rice.

SWEET-AND-SOUR ANIMAL

Prepare animal, cut up, then brown in butter, cover with sliced onions, and pour on the following mixture:

½ cup catsup
¼ cup brown sugar (or granulated)
¼ cup vinegar
Salt and pepper
Onion powder
Garlic powder
Celery salt
Water to thin down

Simmer slowly until tender.

ROAST STUFFED RABBIT

The meat you raise and butcher at home will probably be the meat you cook and eat most often, and for Gretchen and Bob Washburn it's rabbit. Gretchen points out that rabbit is high in protein and low in cholesterol and can be used in any recipe for chicken. "We've had rabbit cooked in cider and breaded and fried, but my family's favorite is good old plain roasted rabbit, where nothing masks the great flavor of the meat," says Gretchen. "The best way to cook rabbit is in a kitchen wood stove like our Kalamazoo," she adds. "But an ordinary kitchen range will do in a pinch!"

3 cups dry bread crumbs for stuffing
1½ cups cornmeal
1 egg
juice of 1 orange (add some pulp, too)
1 cup celery, chopped
1 cup onions, chopped
½ cup walnuts, chopped
pinch of rosemary
pinch of thyme
pinch of parsley
2 rabbits, 2½ to 3 pounds each
cooking oil

Put bread crumbs and cornmeal in bowl and moisten with egg, orange juice, and enough boiling water to make mixture hold together. Add celery, onions, and walnuts. Add herbs. Put stuffing in rabbits and rub them with oil. Place in roasting pan with a little water and roast at 350° for 1 to 1½ hours. Serves 6.

STEAMED VEGETABLES AND LEFTOVER RABBIT

Steam broccoli, carrots, green beans, or any other vegetables. Add cooked rabbit meat and put over hot cooked rice. Top with shredded Cheddar cheese and sprinkle with soy sauce.

MEXICAN RABBIT

This recipe also uses leftover rabbit. Line casserole dish with corn tortillas. Add rabbit meat, green chilis, and shredded Monterey Jack cheese. Top with taco sauce. Add another layer of corn tortillas and repeat layers until dish is filled. Top with more cheese and bake at 350° until bubbly.

MUSKRAT CHILI

Gretchen Washburn warns that this meat is strong in flavor and stringy in texture, but she suggests, "It tastes like what you'd imagine the cowboys ate around the campfire after a long, hard day."

Check to see that musk glands were removed from the muskrat's armpits when animal was skinned. Boil the carcass until the meat falls off the bones and freeze until ready to use.

Put a small amount of oil in pan with a clove of garlic. Brown the clove and remove. Add defrosted meat and brown. Transfer to large pot and cover with boiling water. Add a lot of chili powder, about half as much cumin, and a very little bit of salt and pepper. Add a quart jar of tomatoes to the pot. Cook on low heat, adding a little water if necessary to keep from drying. Cooking all day on a wood stove is best, or 3 hours on modern stove. An hour before serving add 1 or 2 more quarts of tomatoes, 2 cans precooked kidney beans, 1 large chopped onion. Serve with shredded cheese, sour cream, and raw onions.

"I always add a bit of chocolate to chili — it sounds strange, but try it," Gretchen says.

SATURDAY SOUP

There's nothing better for lunch on a cold day or supper after a day of hunting or skiing than a hearty soup, served with a good bread with sweet butter and some cheese, and perhaps some tossed salad — especially if there's a crowd. Let them ladle it out of a steaming tureen and help themselves to crusty hot French bread. It's the perfect lunch for Saturday when the whole family comes home for the weekend, especially because it's best when made ahead. This soup can be made in less time than the recipe here requires, but this recipe is for the working mother who has to cook in the evening to get ready for the weekend.

On Wednesday night take a good meaty shinbone from the freezer and put in a large soup pot with water to cover, 4 quarts or more. Simmer on low heat for a couple of hours, then refrigerate. The next evening remove the fat that has solidified on top of the soup, add chopped celery, onion, and carrots, and give it a couple more hours of simmering, then refrigerate again. On Friday remove the fat again and add peas, mushrooms, and green beans from the freezer, or any other vegetables you choose (mixed frozen vegetables can be used). Put in a can of tomatoes cut up, too, and a couple of handfuls of barley. Also season as you cook, with salt, pepper, a dash of sherry peppers, and any herb you like in soup. Add water if necessary while cooking. You may sample the soup on Friday, but after refrigerating overnight and adding a can of condensed tomato soup while heating again, it's better yet on Saturday.

If you think the soup is good then, however, just wait until Monday when you take out the little bit that's left and find it's so richly jelled that you could turn the bowl over and it wouldn't run out. Reheat for a lunchtime treat.

Before serving the soup, of course, the bone should be removed and the meat cut up and returned to the soup (at the final eating it's almost a stew). But save a little out to slice cold for sandwiches with butter and catsup. Nothing beats a soup-meat sandwich for lunch.

BEST CHICKEN BARBEQUE SAUCE

Most barbecued chicken is smothered with tomato sauce, which tends to hide the more subtle flavor of the meat. This sauce, on the other hand, which is a long-time family favorite, enhances the flavor of the chicken as it cooks, making it more succulent.

⅔ cup butter or margarine
2 tablespoons sugar
1 teaspoon salt
Few grains cayenne
2 tablespoons flour
⅔ cup water or stock
2 tablespoons pickle, chopped (optional)
2 teaspoons Worcestershire sauce
1½ tablespoons lemon juice
¼ cup vinegar
¼ teaspoon Tabasco sauce

Melt butter or margarine, combine dry ingredients and add, stirring until well blended. Remove from heat. Combine remaining ingredients and gradually stir into the butter mixture. Return to heat and cook, stirring constantly, until thick and smooth. Makes about 1⅓ cups, enough for two broiled chickens. Start chicken over hot coals to sear on each side, then begin brushing on sauce (which thickens more if refrigerated until used), continuing as chicken is turned, for about 40 to 45 minutes.

GERMAN CHICKEN WITH INDIAN FLAVOR

Chicken is perhaps the most versatile of meats, and recipes for preparing it are endless. This one is unique in that it comes from Germany but has an Indian flavor. It was translated by Gertraud Stosiek, who brings it from her native country.

1 frying chicken
½ cup butter, melted
1 teaspoon curry
1 teaspoon cinnamon
1 teaspoon cooking oil
½ cup slivered almonds
1 cup orange juice
1 8-ounce jar mango chutney
½ cup raisins
Small can mandarin oranges
1 tablespoon cocktail cherries.

Heat oven to 350°. Cut up chicken, place in baking pan or casserole, and brush with melted butter. Bake 30 minutes. Meanwhile, add curry and cinnamon to oil, heat, and add almonds to toast them. Add orange juice, chutney, and raisins. Salt and sugar may be added to taste. Pour sauce over chicken and bake another 30 minutes. When almost done add drained oranges and cherries. Serve with rice.

WILD GOOSE OR DUCK

Cut up and bone the bird and cut into serving portions. Season and brown in butter. Place in pressure cooker with 1½ cups water or soup stock and cook 20 to 25 minutes at 10 pounds pressure. Add flour and water mixture to thicken gravy. Meat will be tender enough to cut with a fork.

PHEASANT (OR PARTRIDGE) PICCATA

Herb Bergquist teaches junior high school science and inspires a lot of kids to know more about the outdoors, which he enjoys so much himself. Perhaps his favorite sport is bird hunting, and his wife, Flora, who has a nursery school and does a lot of creative cooking besides, has devised many recipes for the game birds Herb brings home.

Debone and skin bird. Cut into small portions and pound thin. Dip pieces in mixture of 2 eggs, 1 tablespoon mayonnaise, then into flour seasoned with salt and pepper. Melt 4 tablespoons butter in skillet over very low heat, add 2 tablespoons lemon juice (more or less according to taste). Add meat and brown lightly, about 2 to 3 minutes on each side (don't let it become crusty, only golden brown). Remove and add 2 cups water, 2 to 3 more tablespoons water, and a chicken bouillon cube. The sauce will thicken a bit. Pour over meat on platter.

"Birds that are only baked, may be dry," explains Flora. "But these are moist and tender." She suggests another recipe, a sort of stew.

Start the same way as in above recipe, but cut meat into small pieces, do not pound. Roll in flour, salt, and pepper. Melt stick of butter and add 4 to 5 cloves of garlic minced fine. Brown only until golden. Meanwhile cook bones to make stock, with celery, an onion, a carrot, parsley, salt, and pepper. Strain and add to pan of drippings (or use 2 cans College Inn chicken broth). Add 4 to 5 chicken bouillon cubes. Simmer to dissolve. Add meat and simmer on very low heat for 1 hour. The sauce will be nicely thickened. Serve with fresh sliced and sautéed mushrooms. Add about 2 tablespoons dry white wine if desired.

GLOSSARY

AITCHBONE. The pelvic bone.

AORTA. The largest artery, runs from heart up along under backbone.

BARROW. A castrated male pig.

BELLY SWEETBREAD. The PANCREAS.

BRISKET. The breast of an animal, or the meat from it.

BUNG. The anus and rectum in male, or anus, rectum, vulva, and vagina in female animals; also casing made from rectum of animal.

CANNON BONE. On the foreleg, the metacarpal bone between knee and ankle; on the rear leg, the metatarsal bone between hock and ankle.

CASING. Animal intestine, plastic, or other material used to stuff sausage or bologna.

CHITTERLINGS. Small intestines of hogs cooked crisp.

COD. The remains of the scrotum still on the animal after castration, usually filled with fat.

CRACKLINGS. Crisp solid matter that remains after hog fat is melted down.

CURE. To let meat "work" or pickle with salt.

DEWCLAWS. Two vestigial hooves on back of ankle (or fetlock) in cattle, hogs, sheep, goats, and deer.

DIAPHRAGM. Fibrous muscular separation between chest cavity and abdominal cavity.

DRY PICKING. Picking feathers from killed fowl without water.

FASCIA. Thin white or colorless membrane separating skin from muscle, and sometimes surrounding muscle.

FELL. Another word for FASCIA.

FEMUR. The long bone from the pelvis to the STIFLE.

GAMBREL. The flexor tendon in back of the cannon bone or Achilles tendon above the HOCK.

GAMBREL JOINT. The hock joint.

GAMBREL STICK. A piece of wood or metal shaped to hook under the gambrels to hang a carcass.

GILT. A young female pig.

GULLET. The esophagus, a tube from the throat to the stomach.

HOCK. The "heel," or tarsal joint, above the dewclaws.

HUMERUS. The long armbone from the shoulder to the elbow.

LIGHTS (LITES). The lungs.

PANCREAS. The digestive gland attached to the small intestine just beyond the stomach.

PERICARDIUM. The "heart sack," a membrane sack surrounding the heart.

PICKLED. Having been preserved in brine (or vinegar).

PICNIC. The cut in pork between the butt and the elbow, containing the HUMERUS.

PLUCK. The lungs, heart, windpipe, and liver; also, to remove feathers.

POTASH KETTLE. A large (50-gallon) cast-iron kettle formerly used to boil potash, now used to dip hogs for hair removal.

SALTED. Having been preserved in brine or salt.

SCAPULA. The shoulder blade.

SHELLY COW. A thin old cow, mostly skin and bone with almost no fat.

SINGLETREE. Also called a whiffletree. A wooden bar with hooks on end and clevis in center, used to hitch a horse by traces to an implement. May be used to hang animal carcasses for butchering.

SKIRT. The DIAPHRAGM.

SPLEEN. A gland in the abdomen that might be confused with the liver — softer, but the same color.

STIFLE. The joint between the FEMUR and TIBIA or shank bone. The same joint as the human knee.

STOCK TANK. A galvanized water tank.

SWEETBREAD. The thymus gland in the neck of veal or very young beef.

THYMUS. A ductless gland in throat of young animals.

TIBIA. The shankbone, or shinbone, from STIFLE to HOCK.

TRACHEA. The windpipe, from larnyx to lungs.

TRIED. Said of fat that has been melted down by boiling into lard.

TABLE OF WEIGHTS AND MEASURES

You may find recipes for sausage, brine, and other foods in various measurements: 1 teaspoon of this, 1 tablespoon of that, 1 pint, 1 quart, or 1 Kg. The following tables will help you to convert to a measure with which you are comfortable.

Approximate Liquid Measure to Nearest Whole Number

1 drop	= 1 minim	
1 teaspoon	= 1 fluid dram (drachm)	= 4 cc
1 dessert spoon	= 2 fluid drams	= 8 cc
1 tablespoon	= 4 drams (½ oz)	= 15 cc
2 tablespoons	= 1 oz	= 30 cc
1 wine glass	= 2 oz	= 60 cc
1 teacup	= 4 oz	= 120 cc
1 glass	= 8 oz (½ pt)	= 240 cc
1 pint	= 16 oz	= 480 cc (approx. 500)
1 quart	= 32 oz, 2 pts	= 960 cc (1,000)
1 gallon	= 128 oz, 8 pts, 4 qts	= 3,840 (4,000)
1 bushel	= 8 gal, 32 qts	
1 milk can	= 10 gal, 40 qts	
1 barrel	= 50–55 gal.	

Metric Weight

1 Kilogram (Kg) = 1,000 grams (Gm)
1 Hectogram (Hg) = 100 grams
1 Dekagram (Dg) = 10 grams
1 decigram (dg) = .1 gram
1 centigram = .01 gram
1 milligram = .001 gram

Approximate Weight Equivalents

1 gram = 15–16 grains (15.432)
4 gram (3.9) = 1 dram
30 gram (31.1) = 1 oz
500 gram (453.6) = 1 lb
1 Kgm = 2.2 lbs (2.2946)

Metric Measure

1 cubic centimeter (cc) = 1 milliliter (ml)
10 cc = 1 deciliter (dl)
100 cc = 1 centiliter (cl)
1,000 cc = 1 liter (L) (1L)
10 L = 1 dekaliter (DL) (10L)
10 K = 1 hectoliter (HL) (100L)
10 HL = 1 kiloliter (KL) (1,000L)

Note that ml or milliliters are used interchangeably with cc or cubic centimeters.

INDEX

Numbers in italics indicate that illustrations or diagrams appear on that page.

Other Books of Interest

The Canning, Freezing, Curing & Smoking of Meat, Fish & Game. Wilbur F. Eastman, Jr. A do-it-yourself reference book for those who want to know how to prepare meat, fish, and game so that it can be stored for future use. Text and illustrations combine to answer all your questions. $6.95. Order #045-4.

Home Sausage Making. Charles Reavis. Sausage that you make yourself is a gastronomic treat, and is free from the nitrates and chemicals found in processed sausage. In this book you'll learn how to make everything from Brockwurst and Bratwurst to Summer sausage and Italian sausage. $10.95. Order #246-5.

Small-Scale Pig Raising. Dirk van Loon. Here you will learn that raising a feeder pig is the best bet for someone with little land who wants to produce the most meat for the smallest investment in time and money. $9.95. Order #136-1.

Raising Poultry the Modern Way. Leonard S. Mercia. How to house, feed, breed, and raise chickens, turkeys and waterfowl. Useful to the experienced person as well as the amateur. Packed with information. $8.95. Order #058-6.

Tan Your Hide! Phyllis Hobson. An excellent source of tanning information. It is a fully illustrated, step-by-step guide to home tanning by nine different methods. Unusually complete. $6.95. Order #101-9.

Country Wisdom Bulletins
(32 pages each)

Build a Smokehouse, A-81. Included are four smokehouse projects (complete with instructions and material lists), and an explanation of the process by which meat and fish are smoked. $1.95. Order #295-3.

The Best Fences, A-92. Planning, construction, the right tools to use, gate construction, and safety for seven different kinds of fences. $1.95. Order #335-6.

These books are available at your bookstore, lawn and garden center, or may be ordered directly from Garden Way Publishing, Department 4412, Schoolhouse Road, Pownal, Vermont 05261. Send for your free mail order catalogue. Please add $2.00 to your order for postage and handling.